I0489262

Water-Quality Characteristics and Trends for Selected Sites At and Near the Idaho National Laboratory, Idaho, 1949–2009

By Roy C. Bartholomay, Linda C. Davis, Jason C. Fisher, Betty J. Tucker, and Flint A. Raben

DOE/ID-22219
Prepared in cooperation with the U.S. Department of Energy

Scientific Investigations Report 2012–5169

U.S. Department of the Interior
U.S. Geological Survey

U.S. Department of the Interior
KEN SALAZAR, Secretary

U.S. Geological Survey
Marcia K. McNutt, Director

U.S. Geological Survey, Reston, Virginia: 2012

For more information on the USGS—the Federal source for science about the Earth, its natural and living resources, natural hazards, and the environment, visit http://www.usgs.gov or call 1–888–ASK–USGS.

For an overview of USGS information products, including maps, imagery, and publications, visit http://www.usgs.gov/pubprod

To order this and other USGS information products, visit http://store.usgs.gov

Suggested citation:
Bartholomay, R.C., Davis, L.C., Fisher, J.C., Tucker, B.J., and Raben, F.A., 2012, Water-quality characteristics and trends for selected sites at and near the Idaho National Laboratory, Idaho, 1949–2009: U.S. Geological Survey Scientific Investigations Report 2012–5169 (DOE/ID 22219), 68 p. plus appendixes.

Contents

Figures

Tables

Conversion Factors, Datums, and Abbreviations and Acronyms

Conversion Factors

Inch/Pound to SI

Multiply	By	To obtain
Length		
foot (ft)	0.3048	meter (m)
mile (mi)	1.609	kilometer (km)
Area		
square mile (mi^2)	2.590	square kilometer (km^2)
Flow rate		
acre-foot per year (acre-ft/yr)	1,233	cubic meter per year (m^3/yr)
Hydraulic conductivity		
foot per day (ft/d)	0.3048	meter per day (m/d)
Hydraulic gradient		
foot per mile (ft/mi)	0.1894	meter per kilometer (m/km)
Transmissivity*		
foot squared per day (ft^2/d)	0.09290	meter squared per day (m^2/d)

SI to Inch/Pound

Multiply	By	To obtain
Volume		
liter (L)	33.82	ounce, fluid (fl. oz)
Mass		
kilogram (kg)	2.205	pound avoirdupois (lb)
Radioactivity		
picocurie per liter (pCi/L)	0.037	Becquerel per liter Bq/L

Temperature in degrees Celsius (°C) may be converted to degrees Fahrenheit (°F) as follows:

$$°F = (1.8 \times °C) + 32.$$

Specific conductance is given in microsiemens per centimeter at 25 degrees Celsius (µS/cm at 25 °C).

Concentrations of chemical constituents in water are given either in milligrams per liter (mg/L) or micrograms per liter (µg/L).

Datums

Vertical coordinate information is referenced to the National Geodetic Vertical Datum of 1929 (NGVD 29).

Horizontal coordinate information is referenced to the North American Datum of 1927 (NAD 27).

Altitude, as used in this report, refers to distance above the vertical datum.

Abbreviations and Acronyms

Abbreviation or Acronym	Definition
ATRC	Advanced Test Reactor Complex (formerly RTC, Reactor Technology Complex, and TRA, Test Reactor Area)
CFCs	chlorofluorocarbons
DOE	U.S. Department of Energy
ESRP	eastern Snake River Plain
E-value	estimated value
INEL	Idaho National Engineering Laboratory (1974–97)
INL	Idaho National Laboratory
INTEC	Idaho Nuclear Technology and Engineering Center
LRL	laboratory reporting level
LT-MDL	long-term method detection level
MCL	maximum contaminant level
NRF	Naval Reactors Facility
NWIS	National Water Information System
NWQL	National Water Quality Laboratory (USGS)
PBF	Power Burst Facility
RESL	Radiological and Environmental Sciences Laboratory (DOE)
QA	quality assurance
RWMC	Radioactive Waste Management Complex
s	sample standard deviation
SDA	Subsurface Disposal Area
TAN	Test Area North
TOC	total organic carbon
USGS	U.S. Geological Survey
VOC	volatile organic compound

Water-Quality Characteristics and Trends for Selected Sites At and Near the Idaho National Laboratory, Idaho, 1949–2009

By Roy C. Bartholomay, Linda C. Davis, Jason C. Fisher, Betty J. Tucker, and Flint A. Raben

Abstract

The U.S. Geological Survey, in cooperation with the U.S. Department of Energy, analyzed water-quality data collected from 67 aquifer wells and 7 surface-water sites at the Idaho National Laboratory (INL) from 1949 through 2009. The data analyzed included major cations, anions, nutrients, trace elements, and total organic carbon. The analyses were performed to examine water-quality trends that might inform future management decisions about the number of wells to sample at the INL and the type of constituents to monitor. Water-quality trends were determined using (1) the nonparametric Kendall's *tau* correlation coefficient, *p*-value, Theil-Sen slope estimator, and summary statistics for uncensored data; and (2) the Kaplan-Meier method for calculating summary statistics, Kendall's *tau* correlation coefficient, *p*-value, and Akritas-Theil-Sen slope estimator for robust linear regression for censored data.

Statistical analyses for chloride concentrations indicate that groundwater influenced by Big Lost River seepage has decreasing chloride trends or, in some cases, has variable chloride concentration changes that correlate with above-average and below-average periods of recharge. Analyses of trends for chloride in water samples from four sites located along the Big Lost River indicate a decreasing trend or no trend for chloride, and chloride concentrations generally are much lower at these four sites than those in the aquifer. Above-average and below-average periods of recharge also affect concentration trends for sodium, sulfate, nitrate, and a few trace elements in several wells. Analyses of trends for constituents in water from several of the wells that is mostly regionally derived groundwater generally indicate increasing trends for chloride, sodium, sulfate, and nitrate concentrations. These increases are attributed to agricultural or other anthropogenic influences on the aquifer upgradient of the INL.

Statistical trends of chemical constituents from several wells near the Naval Reactors Facility may be influenced by wastewater disposal at the facility or by anthropogenic influence from the Little Lost River basin. Groundwater samples from three wells downgradient of the Power Burst Facility area show increasing trends for chloride, nitrate, sodium, and sulfate concentrations. The increases could be caused by wastewater disposal in the Power Burst Facility area.

Some groundwater samples in the southwestern part of the INL and southwest of the INL show concentration trends for chloride and sodium that may be influenced by wastewater disposal. Some of the groundwater samples have decreasing trends that could be attributed to the decreasing concentrations in the wastewater from the late 1970s to 2009. The young fraction of groundwater in many of the wells is more than 20 years old, so samples collected in the early 1990s are more representative of groundwater discharged in the 1960s and 1970s, when concentrations in wastewater were much higher. Groundwater sampled in 2009 would be representative of the lower concentrations of chloride and sodium in wastewater discharged in the late 1980s. Analyses of trends for sodium in several groundwater samples from the central and southern part of the eastern Snake River aquifer show increasing trends. In most cases, however, the sodium concentrations are less than background concentrations measured in the aquifer. Many of the wells are open to larger mixed sections of the aquifer, and the increasing trends may indicate that the long history of wastewater disposal in the central part of the INL is increasing sodium concentrations in the groundwater.

Introduction

The Idaho National Laboratory (INL), operated by the U.S. Department of Energy (DOE), encompasses about 890 mi^2 of the eastern Snake River Plain (ESRP) in southeastern Idaho (fig. 1). The INL was established in 1949 to develop atomic energy, nuclear safety, defense programs, environmental research, and advanced energy concepts. Wastewater disposal sites at the Test Area North (TAN), the Naval Reactors Facility (NRF), the Advanced Test Reactor Complex (ATRC), and the Idaho Nuclear Technology and Engineering Center (INTEC) (fig. 1) have contributed radioactive- and chemical-waste contaminants to the ESRP aquifer. These sites incorporated various wastewater disposal methods, including lined evaporation ponds, unlined infiltration ponds and ditches, drain fields, and injection wells. Waste materials buried in shallow pits and trenches within the Subsurface Disposal Area (SDA) at the Radioactive Waste Management Complex (RWMC) also have contributed contaminants to groundwater.

Since 1949, the U.S. Geological Survey (USGS) has worked in cooperation with the DOE at the INL to define: (1) the quality and availability of water for human consumption, (2) the usability of the water for supporting construction and cooling of facilities and for diluting concentrated waste streams, (3) the location and movement of contaminants in the ESRP aquifer, (4) the sources of recharge to the aquifer, (5) an early detection network for contaminants moving past the INL boundaries, and (6) the processes controlling the origin and distribution of contaminants and naturally occurring constituents in the aquifer (Ackerman and others, 2010).

Since its inception, this water-quality monitoring program at the INL has included a network that once numbered as many as 200 wells. The network of wells has been sampled over the years for tritium, strontium-90, iodine-129, cesium-137, plutonium-238, plutonium-239, -240 (undivided), americium-241, gross alpha- and gross beta-radioactivity, sodium, bromide, chloride, fluoride, sulfate, nitrate, chromium and other trace elements, volatile organic compounds (VOCs), and total organic carbon (TOC) (Bartholomay, 2009; Davis, 2010). Most of the wells in this network were constructed as open boreholes, and many are open to the aquifer throughout their entire depth below the water table.

Purpose and Scope

This report presents an analysis of water-quality data characteristics and trends collected from selected wells and surface-water sites at and near the INL. Water-quality trends are examined to aid future management decisions regarding the number of wells to sample at the INL and the

type of constituents to monitor. The criteria for selecting the sampling sites analyzed in this report were (1) that at least 10 years of data were available for each site, and (2) that the sites represented water that probably was not affected by INL wastewater disposal as related to radionuclide concentrations. Future work will examine water-quality trends in aquifer wells and in perched groundwater known to be affected by wastewater disposal.

Selected constituents were analyzed in water collected from 67 aquifer wells and 7 surface-water sites. The selected constituents were tritium, strontium-90, cesium-137, plutonium-238, plutonium-239, -240 (undivided), americium-241, gross alpha- and beta-particle radioactivity, calcium, magnesium, potassium, silica, sodium, bromide, chloride, fluoride, sulfate, nitrate (as N), orthophosphate (as P), chromium and other trace elements, and total organic carbon. For uncensored data, concentrations of selected constituents were plotted against time to evaluate the data; further trend analyses were performed using the nonparametric Kendall's *tau* correlation coefficient and the Theil-Sen slope estimator; and summary statistics were calculated. For censored data, Kaplan-Meier estimates for calculating summary statistics, Kendall's *tau*, and the Akritas-Theil-Sen slope estimator for robust linear regression were calculated. Because nonparametric methods were used, the assumption of normality was not required.

Geohydrologic Setting

The INL is located on the west-central part of the ESRP. The ESRP is a northeast-trending structural basin about 200 mi long and 50 to 70 mi wide (fig. 1). The basin, bounded by faults on the northwest and by downwarping and faulting on the southeast, has been filled with basaltic lava flows interbedded with terrestrial sediments. The basaltic rocks and sedimentary deposits combine to form the ESRP aquifer, which is the primary source of groundwater on the plain.

The ESRP aquifer is one of the most productive aquifers in the United States (U.S. Geological Survey, 1985, p. 193). Groundwater generally moves from northeast to southwest, and eventually discharges to springs along the Snake River downstream of Twin Falls, Idaho, about 100 mi southwest of the INL (fig. 1). Groundwater moves horizontally through basalt interflow zones and vertically through joints and interfingering edges of interflow zones. Infiltration of surface water, heavy pumpage, geohydrologic conditions, and seasonal fluxes of recharge and discharge locally affect the movement of groundwater (Garabedian, 1986). The ESRP aquifer is recharged primarily from infiltration of applied irrigation water, infiltration of streamflow, groundwater inflow from adjoining mountain drainage basins, and infiltration of precipitation (Ackerman and others, 2006).

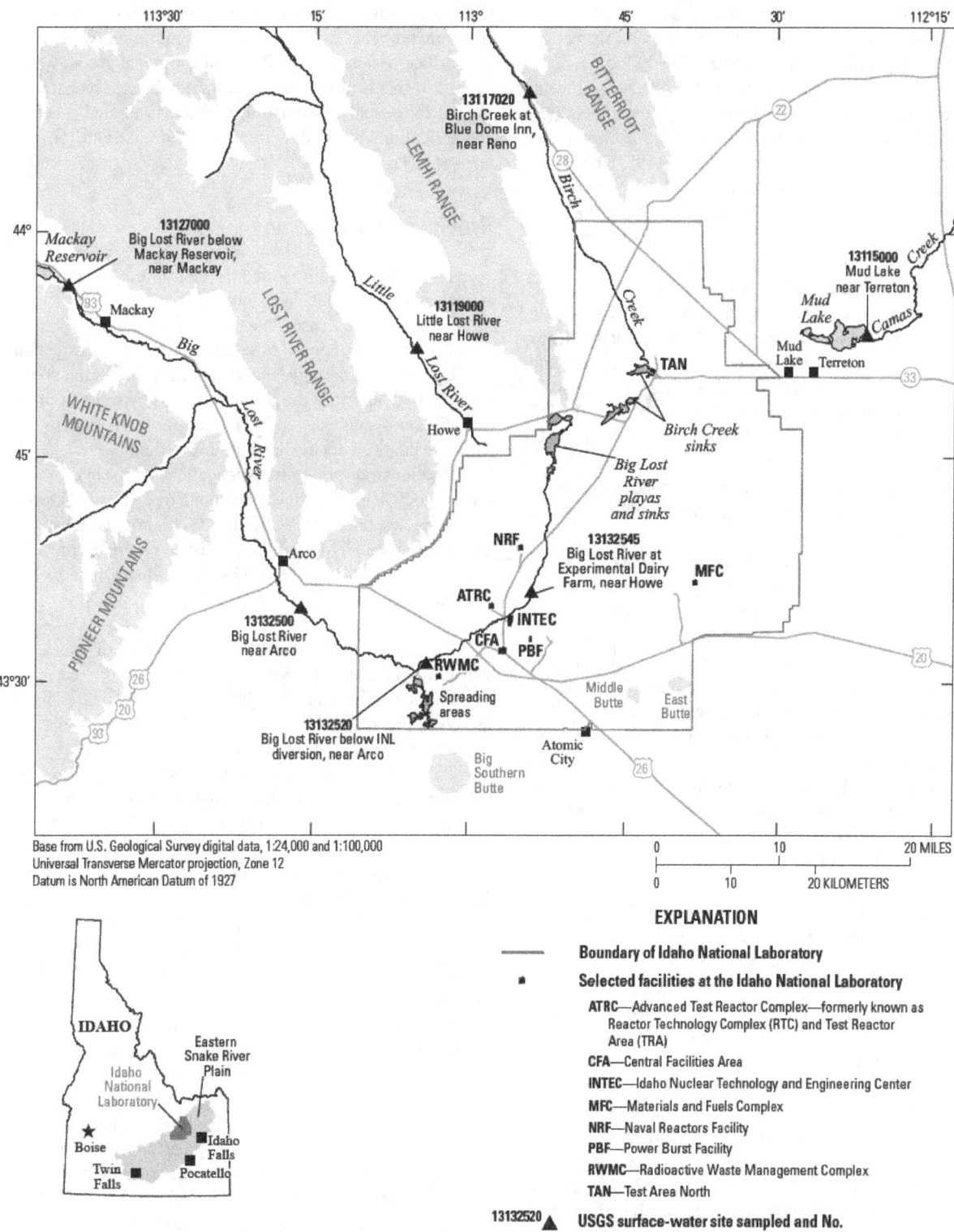

EXPLANATION

— — — — — Boundary of Idaho National Laboratory

▪ Selected facilities at the Idaho National Laboratory

ATRC—Advanced Test Reactor Complex—formerly known as
Reactor Technology Complex (RTC) and Test Reactor
Area (TRA)

CFA—Central Facilities Area

INTEC—Idaho Nuclear Technology and Engineering Center

MFC—Materials and Fuels Complex

NRF—Naval Reactors Facility

PBF—Power Burst Facility

RWMC—Radioactive Waste Management Complex

TAN—Test Area North

13132520 ▲ USGS surface-water site sampled and No.

Figure 1.   Location of surface-water sites and selected facilities, Idaho National Laboratory, Idaho.

At the INL, depth to water in wells completed in the ESRP aquifer ranges from about 200 ft below land surface in the northern part of the INL to more than 900 ft below land surface in the southeastern part of the INL. A significant proportion of the groundwater moves through the upper 200 to 800 ft of basaltic rock (Mann, 1986, p. 21). Ackerman (1991a, p. 30) and Bartholomay and others (1997, table 3) reported transmissivity values for basalt in the upper part of the aquifer ranging from 1.1 to 760,000 ft²/d. The hydraulic gradient at the INL ranges from 2 to 10 ft/mi, with an average of 4 ft/mi (Davis, 2010, fig. 9). Horizontal flow velocities of 2 to 25 ft/d have been calculated based on the movement of various constituents in different areas of the aquifer at and near the INL (Robertson and others, 1974; Mann and Beasley, 1994; Cecil and others, 2000; Plummer and others, 2000; and Busenberg and others, 2001). These flow rates equate to a travel time of about 60 to 700 years for water beneath the INL to travel to springs that discharge at the terminus of the ESRP groundwater flow system. Localized tracer tests at the INL have shown that vertical- and horizontal-transport rates are as high as 60 to 150 ft/d (Nimmo and others, 2002; Duke and others, 2007).

Olmsted (1962), Robertson and others (1974), and Busenberg and others (2001) classified groundwater at the INL based on chemical types derived from dissolution of the rocks and minerals within the recharge source areas. Olmsted's type A water consisted of calcium and magnesium concentrations that constituted at least 85 percent of the cations and bicarbonate constituted at least 70 percent of the anions. Type A water is present in the central and western part of the INL. Type A water is attributed to seepage loss from the Big Lost River and from groundwater underflow from the Big Lost River, Little Lost River, and Birch Creek drainage basins to the west and northwest of the INL (fig. 1) that contain alluvium derived from Paleozoic carbonate rocks from the surrounding mountains.

Olmsted's type B water, which is characterized by higher equivalent fractions of sodium, potassium, fluoride, and silica than type A water, underlies much of the eastern part of the INL and is often referred to as regional water. The groundwater originates from the area northeast of the INL that is composed of a much higher fraction of rhyolitic and andesitic volcanic rocks than mountains surrounding the INL that contribute to Type A water. Busenberg and others (2001) used age dating techniques of chlorofluorocarbons (CFCs), sulfur hexafluoride, and tritium/helium to further classify the regional water at the INL into two types based on the recharge type of the young fraction of groundwater. Water in the southeastern part of the INL represented a binary mixture of old (water greater than 40 and 55 years old that did not contain tritium or CFCs, respectively) regional groundwater

underflow with young water derived from rapid, focused recharge, probably from precipitation infiltration. Water in the northeastern part of the INL is old, regional groundwater underflow that is mixed with local rapid, focused recharge; slow, diffuse areal recharge through the unsaturated zone; and agricultural return flow from the Mud Lake and Terreton areas (fig. 1).

Previous Investigations

Hydrologic conditions and the distribution of selected wastewater constituents in groundwater and perched groundwater are discussed in a series of reports describing the INL. Table 1 summarizes selected previous investigations of the geology, hydrology, and water characteristics at and near the INL, specifies the periods covered by some of those investigations, and lists report citations. Full references for these citations are available in the section, "References Cited." Numerous previous investigations on the hydrology and geology at the INL have been done by INL contractors, state agencies, and the USGS. The USGS provides a list of references and hyperlinks to published reports from its previous INL studies at the USGS INL Project Office Web site, accessed May 7, 2012, at http://id.water.usgs.gov/projects/INL/index.html.

Some qualitative information about trends of selected constituents for selected wells is given in the more recent hydrologic conditions reports listed in table 1, and trend data are given for wastewater constituents from several wells that are affected by wastewater disposal in Davis (2010). Trends for most of the constituents plotted show decreasing concentrations because of the changes in the methods of disposal, dilution, and dispersion in the aquifer, and reductions of amounts of constituents being discharged. One exception is an increasing trend in carbon tetrachloride in a well at the RWMC (Davis, 2010, fig. 29).

Knobel (2006) plotted water-quality data from 1990 through 2003 for 30 wells downgradient of the INTEC to visually compare historical data collected after purging three well volumes with data collected after purging one well volume. Qualitative and quantitative evaluations indicated that the data were comparable for the 30 wells studied.

Concentration trends for selected wells not affected by wastewater disposal at the INL are given for iron, zinc, and nitrate in U.S. Department of Energy (2010). That study indicated that the use of galvanized pipes in some of the wells probably is responsible for elevated zinc concentrations. Concentrations for nitrate were plotted for wells USGS 2, 100, and 101, and the increasing trends were attributed to long-term agricultural practices northeast of the INL.

Table 1. Summary of selected previous investigations on geology, hydrology, and water characteristics of groundwater and perched groundwater, Idaho National Laboratory, Idaho, 1961–2008.

[Modified from Davis (2010). **Summary:** ICPP, Idaho Chemical Processing Plant; INEL, Idaho National Engineering Laboratory; INEEL, Idaho National Engineering and Environmental Laboratory; INL, Idaho National Laboratory; INTEC, Idaho Nuclear Technology and Engineering Center; NRTS, National Reactor Testing Station; RTC, Reactor Technology Complex; RWMC, Radioactive Waste Management Complex]

Reference	Investigation period	Summary
Groundwater		
Jones (1961)		Hydrology of waste disposal at the NRTS, Idaho.
Olmsted (1962)		Chemical and physical character of groundwater at the NRTS, Idaho.
Morris and others (1963, 1964, 1965)		Hydrology of waste disposal at the NRTS, Idaho.
Barraclough and others (1967a)	1965	Hydrology of the NRTS, Idaho.
Barraclough and others (1967b)	1966	Hydrology of the NRTS, Idaho.
Nace and others (1975)		Generalized geologic framework of the NRTS, Idaho.
Robertson and others (1974)		Effects of waste disposal on the geochemistry of groundwater at the NRTS, Idaho.
Barraclough and others (1976)		Hydrology of the solid waste burial ground (now the RWMC).
Barracough and Jenson (1976)	1971–73	Hydrologic data for the Idaho INEL, Idaho.
Barraclough and others (1981)	1974–78	Hydrologic conditions for the INEL, Idaho.
Lewis and Jensen (1985)	1979–81	Hydrologic conditions for the INEL, Idaho.
Pittman and others (1988)	1982–85	Hydrologic conditions for the INEL, Idaho.
Ackerman (1991a)		Analyzed data from 183 aquifer tests conducted in 94 wells to estimate transmissivity of basalts and sedimentary interbeds containing groundwater beneath the INL.
Orr and Cecil (1991)	1986–88	Hydrologic conditions and distribution of selected chemical constituents in water at the INEL, Idaho.
Bartholomay and others (1995)	1989–91	Hydrologic conditions and distribution of selected radiochemical and chemical constituents in water, INEL, Idaho.
Bartholomay and others (1997)	1992–95	Hydrologic conditions and distribution of selected radiochemical and chemical constituents in water, INEL, Idaho.
Bartholomay and others (2000)	1996–98	Hydrologic conditions and distribution of selected constituents in water, INEEL, Idaho.
Davis (2006b)	1999–2001	Hydrologic conditions and distribution of selected radiochemical and chemical constituents in water, INL, Idaho.
Ackerman and others (2006)		Conceptual model of groundwater flow in the eastern Snake River Plain aquifer, INL, with implications for contaminant transport.
Davis (2008)	2002–05	Hydrologic conditions and distribution of selected radiochemical and chemical constituents in groundwater and perched groundwater, INL, Idaho.
Davis (2010)	2006–08	Hydrologic conditions and distribution of selected radiochemical and chemical constituents in groundwater and perched groundwater, INL, Idaho.
Perched groundwater		
Barraclough and others (1967a)	1965	Extent of perched groundwater and distribution of selected wastewater constituents in perched groundwater at the RTC.
Barraclough and others (1967b)	1966	Extent of perched groundwater and distribution of selected wastewater constituents in perched groundwater at the RTC.
Robertson and others (1974)		Analysis of perched groundwater and conditions related to the disposal of wastewater to the subsurface at the INEL.
Barraclough and Jensen (1976)		Extent of perched groundwater and distribution of selected wastewater constituents in perched groundwater at the RTC.
Robertson (1976)		Numerical model simulating flow and transport of chemical and radionuclide constituents through perched water at the RTC.
Barraclough and others (1981)	1974–78	Hydrologic conditions for the INEL, Idaho.
Lewis and Jensen (1985)	1979–81	Hydrologic conditions for the INEL, Idaho.
Pittman and others (1988)	1982–85	Hydrologic conditions for the INEL, Idaho.
Hull (1989)		Conceptual model that described migration pathways for wastewater and constituents from the radioactive-waste infiltration ponds at the RTC.
Anderson and Lewis (1989)		Correlation of drill cores and geophysical logs to describe a sequence of basalt flows and sedimentary interbeds in the unsaturated zones underlying the RWMC.

Table 1.   Summary of selected previous investigations on geology, hydrology, and water characteristics of groundwater and perched groundwater, Idaho National Laboratory, Idaho, 1961–2008.—Continued

[Modified from Davis (2010). **Summary:** ICPP, Idaho Chemical Processing Plant; INEL, Idaho National Engineering Laboratory; INEEL, Idaho National Engineering and Environmental Laboratory; INL, Idaho National Laboratory; INTEC, Idaho Nuclear Technology and Engineering Center; NRTS, National Reactor Testing Station; RTC, Reactor Technology Complex; RWMC, Radioactive Waste Management Complex]

Reference	Investigation period	Summary
	Perched groundwater—Continued	
Anderson (1991)		Correlation of drill cores and geophysical logs to describe a sequence of basalt flows and sedimentary interbeds in the unsaturated zones underlying the RTC and INTEC.
Ackerman (1991b)		Analyzed data from 43 aquifer tests conducted in 22 wells to estimate transmissivity of basalts and sedimentary interbeds containing perched groundwater beneath the RTC and INTEC.
Cecil and others (1991)	1986–88	Mechanisms for formation of perched water at the RTC, ICPP, and RWMC, INEL, Idaho; distribution of chemical and radiochemical constituents in perched water at the RTC, ICPP, and RWMC.
Tucker and Orr (1998)		Hydrologic conditions and distribution of selected radiochemical and chemical constituents in perched groundwater, INEL, Idaho.
Bartholomay (1998)	1992–95	Hydrologic conditions and distribution of selected radiochemical and chemical constituents in perched groundwater, INEL, Idaho.
Orr (1999)		Transient numerical simulation to evaluate a conceptual model of flow through perched water beneath wastewater infiltration ponds at the RTC.
Bartholomay and Tucker (2000)	1996–98	Hydrologic conditions and distribution of selected radiochemical and chemical constituents in perched groundwater, INEEL, Idaho.
Davis (2006a)	1999–2001	Hydrologic conditions and distribution of selected radiochemical and chemical constituents in perched groundwater, INL, Idaho.
Davis (2008)	2002–05	Hydrologic conditions and distribution of selected radiochemical and chemical constituents in groundwater and perched groundwater, INL, Idaho.
Davis (2010)	2006–08	Hydrologic conditions and distribution of selected radiochemical and chemical constituents in groundwater and perched groundwater, INL, Idaho.

Methods

Sample Collection and Analyses

Water samples analyzed for this study were collected at seven surface-water sites (fig. 1) and 67 wells (fig. 2). Since 1989, water samples have been analyzed for chemical constituents at the USGS National Water Quality Laboratory (NWQL) in Lakewood, Colorado. Prior to 1989, water samples were analyzed by various laboratories for chloride, chromium, sodium, and nitrate (Wegner, 1989). Water samples have been analyzed for radiochemical constituents at the DOE Radiological and Environmental Sciences Laboratory (RESL) at the INL since samples were first collected. Many of the samples collected in the 1950s, 1960s, and 1970s were collected during or immediately after cable-drilling, and some of the samples had a high probability of containing impurities introduced by the drilling (Robertson and others, 1974, appendix B). Some chemical and radiochemical analyses for special studies have been done at the NWQL and by contract laboratories throughout the history of the USGS monitoring program at the INL, but those data were not considered in this study because of the variability of method detection levels and the analytical methods used. Until 2008, the RESL reported an uncertainty of $2s$, where s is the sample standard deviation, for water samples analyzed for radionuclides, and data were entered into the USGS National Water Information System (NWIS) database with that uncertainty. In 2008, the RESL began reporting the uncertainty as $1s$, and data are now entered into NWIS with the $1s$ uncertainty. Analytical uncertainties in this report are reported as $1s$ for consistency.

Methods used to sample and analyze for selected constituents generally follow the guidelines established by the USGS (Goerlitz and Brown, 1972; Stevens and others, 1975; Wood, 1976; Thatcher and others, 1977; Claassen, 1982; Wershaw and others, 1987; Fishman and Friedman, 1989; Faires, 1993; Fishman, 1993; and U.S. Geological Survey, variously dated). Water samples were collected according to a quality-assurance plan for water-quality activities conducted by personnel at the USGS INL Project Office. The plan was finalized in June 1989 and revised in March 1992, in 1996 (Mann, 1996), in 2003 (Bartholomay and others, 2003), and in 2008 (Knobel and others, 2008). The plan is available for inspection at the USGS INL Project Office.

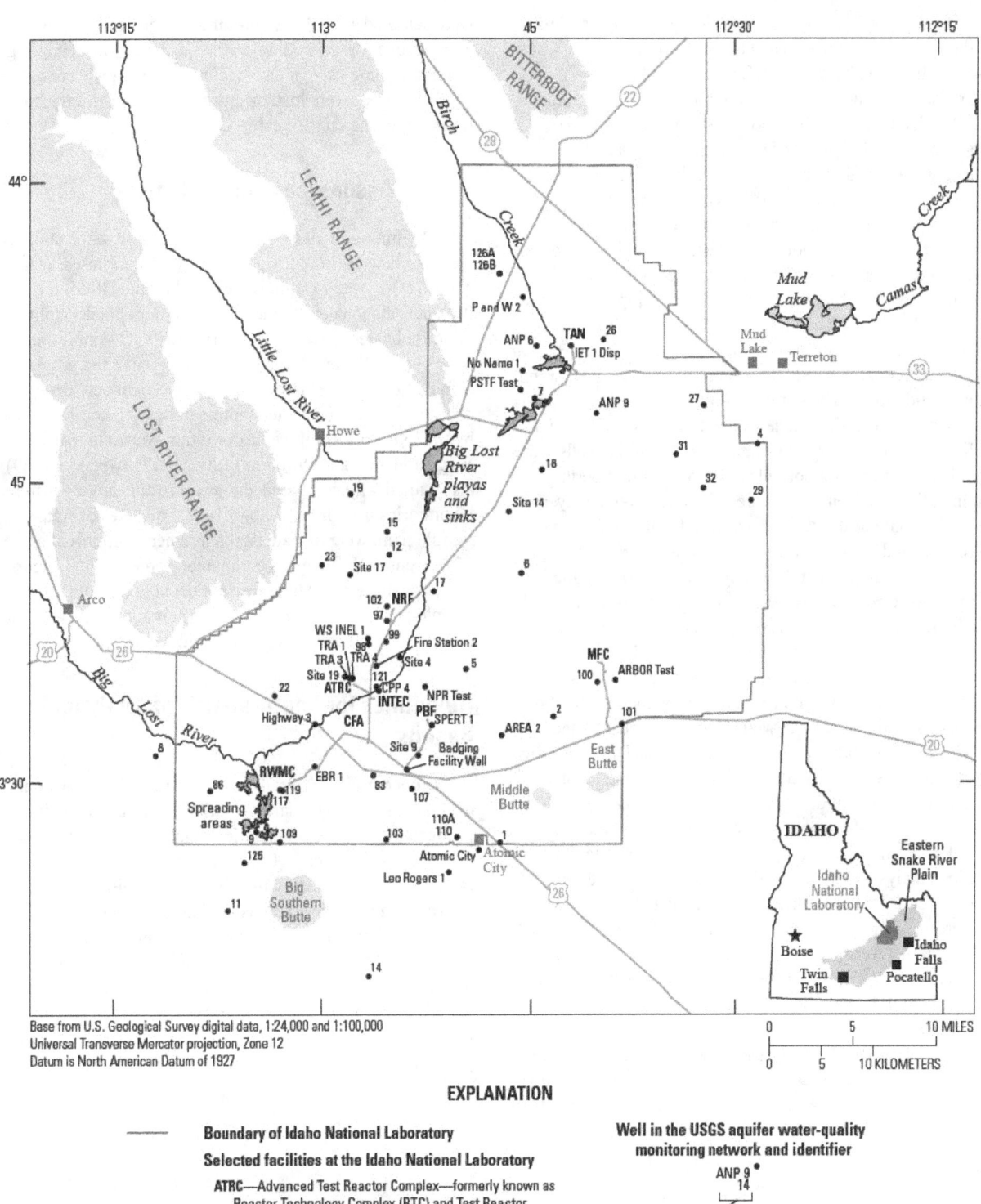

Figure 2.   Location of wells at and near the Idaho National Laboratory, Idaho.

Field processing of water samples differed depending on the constituents for which analyses were requested. Water samples analyzed by the NWQL were placed in containers and preserved in accordance with laboratory requirements specified by Timme (1995) and Knobel and others (2008, appendix A). Containers and preservatives for samples collected since 1989 were supplied by the NWQL and were processed using a rigorous quality-control procedure (Pritt, 1989, p. 75) to minimize sample contamination. Water samples requiring filtration were filtered through a disposable 0.45-μm filter cartridge that was pre-rinsed with at least 1 L of deionized water or sample water. Water samples analyzed by the RESL were placed in containers and preserved in accordance with laboratory requirements specified by Bodnar and Percival (1982), U.S. Department of Energy (1995), and Knobel and others (2008, appendix A).

Sample collection methods varied for several of the wells during the history of sampling. Many of the wells were sampled using either a portable thief sampler or a portable pump until permanent pumps were installed in the early 1990s (table 2). Some of the samples collected with thief samplers were collected at different depths in the aquifer during the same sampling event. When the depths at which thief samples were collected were known, the data from the depth similar to the depth the current pump is set to were used in the analyses. After pumps were installed, wells were purged for at least three well volumes prior to sample collection until October 2003, when procedures were changed to allow sample collection after one well volume was purged and readings were stable for water temperature, pH, and specific conductance. Studies by Bartholomay (1993) and Knobel (2006) indicated that different purge rates used at the INL did not affect the analytical results for the respective studies.

The methods used to preserve samples collected for nutrient analyses also changed over time. Until October 1994, samples collected for nutrients were analyzed at the NWQL and preserved with mercuric chloride and chilling. After that date, samples were preserved only by chilling. Bartholomay and Williams (1996) compared preservation methods, and their results indicated that data were comparable between the two methods.

Sample collection frequency varied for all sites used in this study. Since 2003, all sites have been sampled annually, but prior to that timeframe, wells and surface-water sites were sampled annually, semi-annually, quarterly, or even more frequently depending on the purpose of the sampling program. Some gaps in data occur when pumps were out for repair, samples were lost, or program changes did not call for sampling of the constituent in question.

Quality Assurance/Quality Control

Beginning in 1980, about 10 percent of water samples were collected for quality assurance (QA) purposes. Quality control (QC) water samples collected by the USGS INL Project Office generally include equipment blanks, splits, and blind replicates; however, other types of QC samples also have been collected throughout the history of the program. Comparative studies to determine agreement between analytical results for water-sample pairs by laboratories used by the INL Project Office QA program were summarized by Wegner (1989) and Williams (1996, 1997). Wegner (1989) also statistically compared analytical results among different laboratories used from 1980 to 1988. Analyses of water-sample pairs were in statistical agreement for more than 95 percent of the samples compared. Some outliers occurred; in cases where replicates were collected, data from the samples with the most reasonable result compared with the long-term trend were used in the trend analyses.

Guidelines for Interpretation of Analytical Results

Concentrations of radionuclides are reported with an estimated sample standard deviation, s, which is obtained by propagating sources of analytical uncertainty in measurements. McCurdy and others (2008) provided details on interpreting radiological data used by the USGS. The guidelines for interpreting analytical results are based on an extension of a method proposed by Currie (1984) and are given in Davis (2010). In this report, radionuclide concentrations of less than $3s$ are considered to be less than the "reporting level." The reporting level should not be confused with the analytical method detection limit, which is based on laboratory procedures.

Table 2.   Site information for sampling sites at and near the Idaho National Laboratory, Idaho.

[**Local name:** Local well identifier used in this study. **Site identifier:** Unique numerical identifiers used to access well data (http://waterdata.usgs.gov/nwis). Location of sampling sites are shown in figures 1 and 2. **Well depth:** ft bls, feet below land surface. NA, not applicable]

Sampling sites				
Local name	Site identifier	Well depth (ft bls)	Period of record	Pump installation date
Wells				
ANP 6	435152112443101	305	1956–2009	01-10-1986
ANP 9	434856112400001	322	1956–2009	04-12-1994
ARBOR Test	433509112384801	790	1957–2009	[1]10-01-1988
AREA 2	433223112470201	877	1960–2009	08-07-1990
Atomic City	432638112484101	639	1953–2009	[1]1953
Badging Facility Well	433041112535101	644	1985–2009	[1]1985
CPP 4	433440112554401	700	1983–2009	[1]1983
EBR 1	433051113002601	1,075	1949–2009	[1]1949
Fire Station 2	433548112562301	510	1957–1996	12-04-1957
Highway 3	433256113002501	750	1977–2009	[1]1977
IET 1 Disp	435153112420501	242	1953–2009	01-13-1986
Leo Rogers 1	432533112504901	720	1970–2009	[1]1970
No Name 1	435038112453401	550	1984–2009	07-31-1990
NPR Test	433449112523101	600	1986–2009	01-23-1986
P and W 2	435419112453101	386	1957–2009	01-10-1986
PSTF Test	434941112454201	319	1957–2009	07-30-1990
Site 4	433617112542001	495	1957–2009	[1]1957
Site 9	433123112530101	1,057	1969–2009	08-31-1990
Site 14	434334112463101	717	1956–2009	10-01-1975
Site 17	434027112575701	600	1977–2009	09-06-1990
Site 19	433522112582101	860	1960–2009	01-16-1986
SPERT 1	433252112520301	653	1956–2009	[1]1956
TRA 1	433521112573801	600	1950–2009	[1]1950
TRA 3	433522112573501	602	1972–2009	[1]1972
TRA 4	433521112574201	965	1972–2009	[1]1972
USGS 1	432700112470801	636	1949–2009	07-30-1990
USGS 2	433320112432301	699	1950–2009	07-28-1990
USGS 4	434657112282201	553	1950–2009	07-31-1990
USGS 5	433543112493801	494	1950–2009	09-08-1990
USGS 6	434031112453701	620	1950–2009	09-05-1990
USGS 7	434915112443901	903	1950–2009	07-11-1990
USGS 8	433312113115801	812	1950–2009	07-10-1990
USGS 9	432740113044501	654	1951–2009	07-30-1987
USGS 11	432336113064201	704	1950–2009	09-12-1989
USGS 12	434126112550701	563	1950–2009	05-10-1990
USGS 14	432019112563201	752	1951–2009	09-11-1989
USGS 15	434234112551701	610	1952–2009	01-11-1990
USGS 17	433937112515401	498	1951–2009	09-08-1989
USGS 18	434540112440901	329	1951–2009	08-29-1990
USGS 19	434426112575701	399	1951–2009	09-07-1990
USGS 22	433422113031701	657	1952–2009	07-27-1990

Table 2. Site information for sampling sites at and near the Idaho National Laboratory, Idaho.—Continued

[**Local name:** Local well identifier used in this study. **Site identifier:** Unique numerical identifiers used to access well data (http://waterdata.usgs.gov/nwis). Location of sampling sites are shown in figures 1 and 2. **Well depth:** ft bls, feet land surface. NA, not applicable]

Sampling sites				
Local name	Site identifier	Well depth (ft bls)	Period of record	Pump installation date
Wells–Continued				
USGS 23	434055112595901	458	1952–2009	08-29-1990
USGS 26	435212112394001	266	1952–2009	07-31-1990
USGS 27	434851112321801	312	1952–2009	08-28-1990
USGS 29	434407112285101	426	1953–2009	08-28-1990
USGS 31	434625112342101	428	1953–2009	08-24-1990
USGS 32	434444112322101	392	1953–2009	08-24-1990
USGS 83	433023112561501	752	1965–2009	[1]08–1980
USGS 86	432935113080001	691	1971–2009	08-03-1987
USGS 97	433807112551501	510	1972–2009	01-22-1986
USGS 98	433657112563601	508	1973–2009	01-21-1986
USGS 99	433705112552101	440	1975–2009	01-23-1986
USGS 100	433503112400701	750	1975–2009	01-28-1986
USGS 101	433255112381801	842	1975–2009	01-29-1986
USGS 102	433853112551601	445	1990–2009	05-09-1990
USGS 103	432714112560701	760	1980–2009	[1]11–1986
USGS 107	432942112532801	690	1980–2009	11-21-1983
USGS 109	432701113025601	800	1980–2009	07-30-1987
USGS 110	432717112501501	780	1980–1992	11–1983
USGS 110A	432717112501502	644	1995–2010	10-24-1995
USGS 117	432955113025901	655	1987–2009	10-13-1987
USGS 119	432945113023401	705	1987–2009	10-24-1987
USGS 121	433450112560301	475	1991–2009	[1]1991
USGS 125	432602113052801	774	1995–2009	04-26-1995
USGS 126 A	435529112471301	648	2000–2009	10-25-2000
USGS 126 B	435529112471401	472	2000–2009	11-21-2000
WS INEL 1	433716112563601	490	1981–2009	01-18-1986
Surface-water sites				
Big Lost River at Experimental Dairy Farm, near Howe	13132545	NA	1981–2009	NA
Big Lost River below INL diversion, near Arco	13132520	NA	1981–2009	NA
Big Lost River below Mackay Reservoir, near Mackay	13127000	NA	1985–2009	NA
Big Lost River near Arco	13132500	NA	1965–2009	NA
Birch Creek at Blue Dome Inn, near Reno	13117020	NA	1970–2009	NA
Little Lost River near Howe	13119000	NA	1965–2009	NA
Mud Lake near Terreton	13115000	NA	1965–2009	NA

[1]Estimated date based on month or year of installation.

Concentrations of inorganic and organic constituents are reported with reference to laboratory reporting levels (LRLs). Childress and others (1999) provided details about the approach used by the USGS regarding detection limits and reporting levels. The method detection limit is the minimum concentration of a substance that can be measured and reported with 99 percent confidence that the concentration is greater than zero. The LRL is the concentration at which the false negative error rate is minimized to be no more than 1 percent of the reported results. The LRL generally is equal to twice the yearly determined long-term method detection level (LT-MDL), which is a detection level derived by determining the standard deviation of a minimum of 24 MDL spike-sample measurements over an extended time period. The LT-MDL controls false positive error; it is the concentration at which the false positive risk is minimized to be no more than 1 percent of the reported values (Childress and others, 1999). These reporting levels may be described as preliminary for a developmental method if the levels have been based on a small number of analytical results. These levels also may vary from sample to sample for the same constituent and the same method if matrix effects or other factors arise that interfere with the analysis. Concentrations measured between the LT-MDL and the LRL are described as estimated values. For most of the constituents in this report, reported concentrations generally are greater than LRLs, but some concentrations are given as less than the LRL, and some concentrations are estimated.

Statistical Data Analysis Methods

Water-quality data collected for the monitoring program were analyzed using the summary statistics of mean, median, minimum, maximum, and standard deviation. The data were processed using custom computer scripts developed in the R programming language (R Development Core Team, 2011). All R functions written for this report are stored in an R-package called "Trends" (Fisher and Davis, 2011); function documentation for this package is given in appendix A. The "Trends" package identifies statistical trends in water-quality data for multiple constituents and sample sites using nonparametric statistical tests for both censored data (values reported as less than a LRL) and uncensored data. Nonparametric tests were used for INL data because no

specific distributions of the data were expected. This package also calculates summary statistics for multiple constituents, and automates plotting the data and trend lines for uncensored data. Because nonparametric tests were used, the assumption of normality is not required.

For uncensored data, the Theil-Sen slope estimator—the median of all possible slopes between pairs of data (Helsel, 2005)—and Kendall's *tau* correlation coefficient were calculated. The R functions used to calculate these statistics were developed by Wilcox (2012) and are included in the section, "Trends." Summary statistics for uncensored data are shown in table 3 (at back of report).

Censored data were analyzed using functions provided in the R Package "NADA" (Nondetects and Data Analysis) (Lee, 2012), which were called from "Trends." The "NADA" package uses methods described by Helsel (2005) to deal with statistics for censored environmental data. The summary statistics based on the Kaplan-Meier method, the p-value for the test for significance of Kendall's *tau* correlation coefficient, and the slope of the associated Akritas-Theil-Sen nonparametric regression line (Helsel, 2005, p. 212) are given in table 4 (at back of report). The Akritas-Theil-Sen slope estimator calculates a slope that produces a zero value for *tau* (Helsel, 2005).

For datasets with estimated (E) concentrations, the E-value was treated the same as a concentration for an uncensored value. Summary statistics were not calculated for constituents that had all concentrations less than the reporting level.

For both uncensored and censored datasets, the test statistic used was Kendall's *tau* correlation coefficient. The null hypothesis was that no correlation existed between time and concentration; the alternate hypothesis was that time and concentration were correlated. A significance level of 0.05 was chosen to determine whether the result of the test for significance of Kendalls's *tau* correlation coefficient was statistically significant. A two-sided *p*-value that was less than or equal to 0.05 indicated that there was a statistically significant correlation in the data, and the null hypothesis was rejected. The sign of the slope indicated whether there was an increasing or decreasing trend if a significant correlation existed. *P*-values greater than 0.05 indicated that there was no statistically significant correlation between time and concentration, and the null hypothesis was accepted.

Water-Quality Characteristics and Trends

Concentrations of constituents reported in this study represent water samples collected during different time intervals, starting in 1949 at well USGS 1, and continuing at this and other sites through 2009. Data used for selected plots and trend analyses (appendixes B–E) included all data that was in the NWIS database as of August 2010, except for some outlier data that were removed because it did not correlate with field-specific conductance or because replicate values represented more reasonable data. Data analyzed by the NWQL is routinely uploaded to the NWIS database; however, data from INL laboratories were manually entered from paper copies until 2009. Some data (especially prior to 1990) may still need to be entered and checked for some wells used in this study. Trends were initially analyzed for entire datasets, but analyses of some wells seemed to indicate that pump installation created a change in the concentrations of some constituents. Therefore, trends were determined for the period of record when a pump was installed. Some well pumps were installed for the entire period of record, but trends were analyzed only for data collected from 1980 to 2009 for those wells because some of the earlier data appeared suspect; 1980 represented the start of quality-assurance data collection to support data validity. Trends in surface-water constituent concentrations were determined for the entire period of record. The trends described in the report were based on statistical methods previously described. For some datasets, the slope of the line or the visual appearance of the data based on scale make it appear like there is a trend, when statistically speaking, there is no trend.

Wells were completed at various depths in the aquifer and with different well completions (for example, single and multiple screened intervals and open boreholes). Physical properties of water measured during sampling events included specific conductance, water temperature, and pH.

Samples collected from sites included in this study were analyzed for some combination of radiochemical and chemical constituents tritium, strontium-90, cesium-137, plutonium-238, plutonium-239, -240 (undivided), americium-241, gross alpha- and beta-particle radioactivity, calcium, magnesium, potassium, silica, sodium, bromide, chloride, fluoride, sulfate, nitrate (as N), orthophosphate (as P), chromium and other trace elements, and total organic carbon. Other constituents have been sampled at some of the wells and surface-water sites, but data were not used because concentrations were either less than the reporting levels (in the case of volatile organic compounds) or because data were insufficient to determine a statistically significant trend.

pH, Specific Conductance, and Temperature.—Field measurements of pH, specific conductance, and water temperature were measured in all 67 aquifer wells and 7 surface-water sites used for this study (figs. 1 and 2). Summary statistics for the field measurements for the entire period of record at each surface-water site, and the period of record used to calculate trends in each well are given in table 5 (at back of report). Plots of analyses for pH, specific conductance, and water temperature for the entire datasets are shown in appendix B.

Analyses of trends for pH indicate that some sites have statistically significant trends based on the p-values calculated (table 5). The pH is an indicator of hydrogen ion activity, and is collected to understand the acid-base properties of water. Many factors affect the degree of precision of pH readings including careful attention to the electrode maintenance, buffer solutions, temperature corrections, instrument electronics, use of different meter brands, and collection methods. Because of all the variability of these factors throughout the history of sample collection, it is difficult to specifically assess whether or not the trends in the pH are due to changing aquifer conditions or other considerations. Four wells (ANP 9, PSTF Test, USGS 18, and USGS 97) appear to have been affected by the change in methods and meter type between April and October 2004 (appendix B). Measurements of pH were taken by a single parameter pH meter by placing water in a cup and submerging the probe prior to October 2004 and have since been taken from a flow through chamber with a multiple parameter instrument. Site 9 was affected by the change of sample method from thief sampling to pumping in 1990.

Specific conductance is a measure of the electrical conductivity of water and is proportional to the quantities of dissolved chemical constituents in the water so trends in this measurement should be similar to the trends for chloride, sulfate, and sodium concentrations. Analyses of specific conductance in a few wells (Site 9, USGS 2, 9, 14, and 17) show a discernible change (appendix B) when the sampling method changed from collecting a thief sample to installation of a dedicated pump in the well around 1990. The pump sample represents a mixture of all water in the open interval of the well versus the thief sample that was collected from a discrete depth, and the marked change is an indication of different water types present in the aquifer at the well locations. Well USGS 98 shows a discrete decrease in specific conductance when the pump was lowered in 2005. Several wells (Badging Facility Well, CPP 4, PSTF Test, TRA 3 and 4, USGS 17, 27, 117, and 119) appear to have discernible changes when the collection method changed from taking readings in a cup to taking them from a flow through chamber. Figure 3 shows specific conductance trends by well for the period of record used for trend analyses. Analyses of trends generally show similarities to trends of chemical constituents. Trends for individual chemical constituents are discussed later in this section.

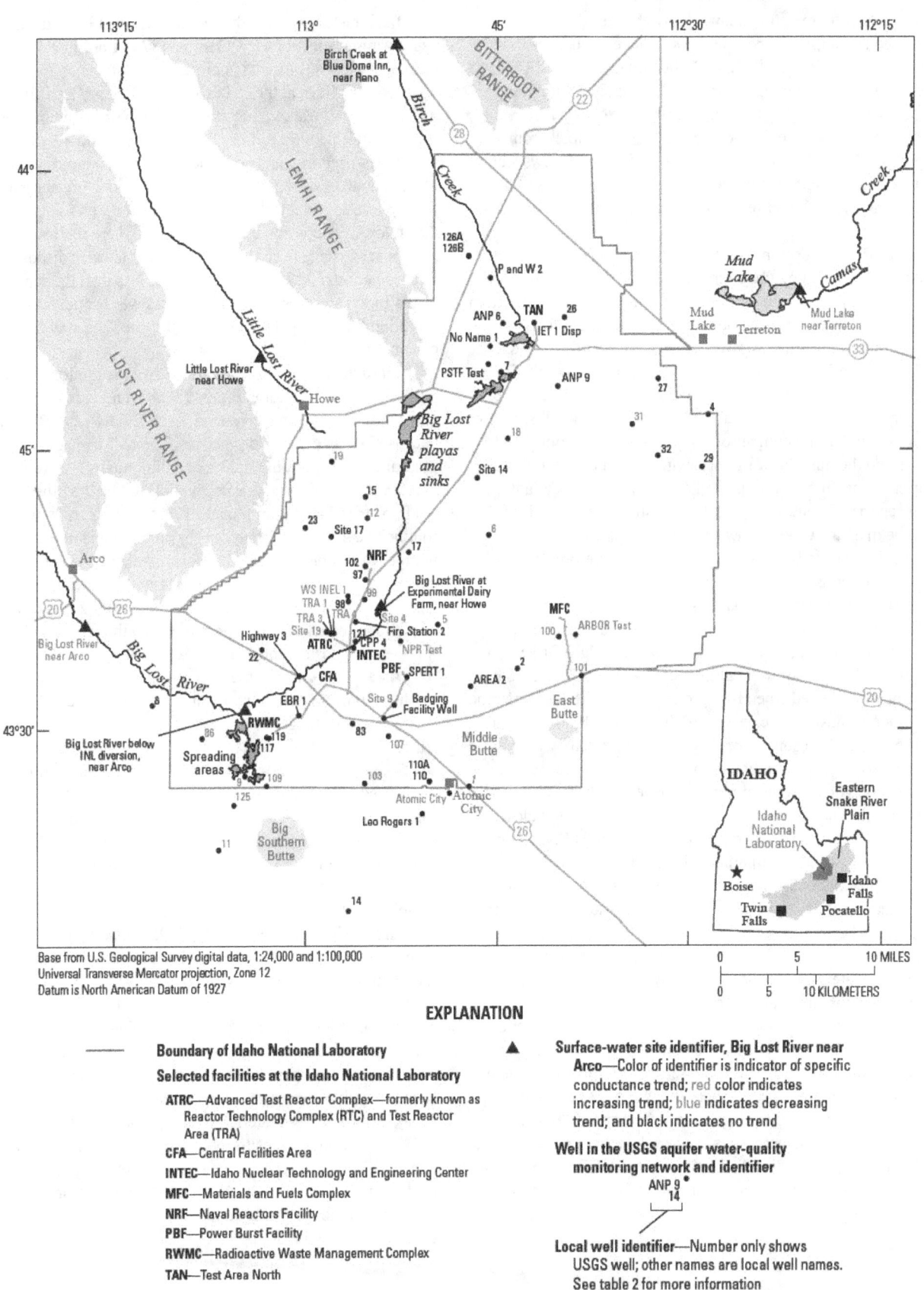

Figure 3. Areal distribution of specific conductance trends in water from selected wells and surface-water sites at and near the Idaho National Laboratory, Idaho.

Water temperature shows a few changes that may be related to sample methods and not to general changes in the aquifer condition. Wells ANP 9, Site 9, USGS 7, and 17 all show a change in temperature that corresponds with pump installation (appendix B, table 2). Because of this factor, trend analyses were done on datasets from samples from pumped wells for the 67 aquifer wells. Figure 4 shows water temperature trends by site for the period of record used for trend analyses. Water temperatures at most of the sites show no trend.

Tritium.—Tritium is a radioactive isotope of hydrogen that is formed in nature by interactions of cosmic rays with gases in the upper atmosphere (Orr and others, 1991). Tritium also is produced in thermonuclear detonations, and it has been discharged in wastewater at several facilities at the INL from the early 1950s to the present. Tritium has a half-life of 12.3 years (Walker and others, 1989, p. 20). Samples were routinely collected and analyzed for tritium from all sites used in this study. The number of samples collected from each site and the number of samples with tritium concentrations greater than the reporting level are shown in table 6 (at back of report). The sites selected for this study were considered to be in areas where the water does not appear to have been influenced by wastewater disposal; therefore, tritium concentrations are less than the reporting levels for almost all samples and no trend analyses were done. A few samples had spurious concentrations that were equal to or greater than the reporting level, or were extremely high outliers. In almost all cases for the high outliers, the results were from samples collected when background concentrations were much higher because of nuclear testing (in the late 1950s and early 1960s), so the results may have been influenced by the higher background concentrations. Some of the more recent samples with analytical results at reporting levels of $3s$ could be due to statistical fluctuations because the counting data are close to background concentrations (Guy Backstrom, U.S. Department of Energy, written commun., July 8, 2009).

Strontium-90.—Strontium-90 is a fission product of nuclear weapons tests, and is present in wastewater discharged at several facilities at the INL (Davis, 2010). Strontium-90 has a half-life of 29.1 years (Walker and others, 1989, p. 29). Water samples have routinely been collected and analyzed for strontium-90 from 25 of the wells used in this study. The number of samples collected and the number of samples with concentrations greater than the reporting level are shown in table 6. The wells selected for this study were considered to be in areas where the water does not appear to have been influenced by wastewater disposal. Because strontium-90 concentrations are less than the reporting levels for all samples, no trend analyses were done. The USGS may discontinue sampling for strontium-90 for some of these sites.

Cesium-137.—The radionuclide cesium-137 is identified using gamma spectrometry. Cesium-137 is a product of nuclear weapons tests, and is present in wastewater discharged at several facilities at the INL. Cesium-137 has a half-life of 30.17 years (Walker and others, 1989). Gross gamma has routinely been collected from 37 of the wells and 4 surface-water sites used in this study. The number of samples collected and the number of samples with concentrations greater than the reporting level are shown in table 6. The wells selected for this study were considered to be in areas where the water does not appear to have been influenced by wastewater disposal. Because cesium-137 concentrations are less than the reporting levels for all samples, no trend analyses were done. Future sampling for cesium-137 for some of these sites may be discontinued.

Plutonium.—In 1974, the USGS began monitoring plutonium-238 and plutonium-239, -240 (undivided) in water from selected wells around TAN, INTEC, and RWMC because of waste disposal practices. The half-lives of plutonium-238, plutonium-239, and plutonium-240 are 87.7, 24,100, and 6,560 years, respectively (Walker and others, 1989, p. 46). Water from four wells used in this study was routinely sampled and analyzed for plutonium isotopes. The number of samples collected and the number of samples with concentrations greater than the reporting level are shown in table 6. Because plutonium-238 and plutonium-239, -240 (undivided) concentrations were less than the reporting levels for all samples, no trend analyses were done.

Americium-241.—Americium-241 is a decay product of plutonium-241. Plutonium isotopes have been detected in wastewater discharged to the ESRP aquifer at the INL and in wastes buried at the RWMC. The half-life of americium-241 is 432.7 years (Walker and others, 1989, p. 46). Water from four wells used in this study was routinely sampled and analyzed for plutonium isotopes. The number of samples collected and the number of samples with concentrations greater than the reporting level are shown in table 6. Because americium-241 concentrations were less than the reporting levels for all samples, no trend analyses were done.

Gross Alpha- and Beta-Particle Radioactivity.—Gross alpha- and beta-particle radioactivity is a measure of the total radioactivity emitted as alpha and beta particles during the radioactive decay process. The radioactivity usually is reported as if it occurred as one radionuclide. Gross alpha and beta measurements are used to screen for radioactivity in the aquifer as a possible indicator of groundwater contamination, but measurable concentrations also can occur from the natural decay of radioactive material in aquifer material. In 2008, the RESL increased the sensitivity of the analyses, and as a result, concentrations greater than the reporting level for natural background concentrations in the aquifer were sometimes measured, although those same concentrations may not have been reported in the past.

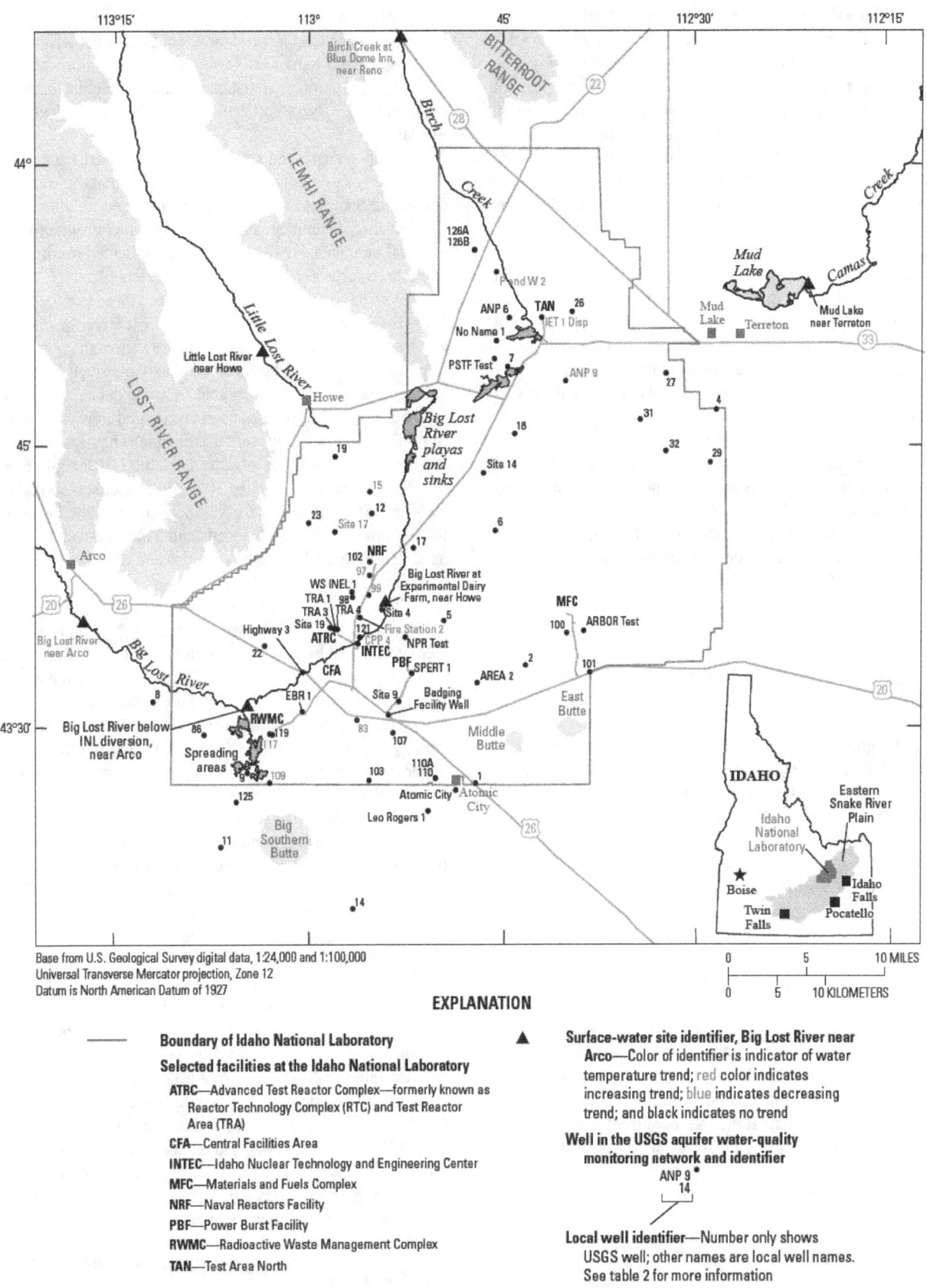

Figure 4. Areal distribution of water temperature trends in water from selected wells and surface-water sites at and near the Idaho National Laboratory, Idaho.

Water samples have routinely been collected and analyzed for gross alpha- and beta-particle activity from 35 of the wells and 4 surface-water sites used in this study. The number of samples collected and the number of samples with concentrations greater than the reporting level are shown in table 6. The wells selected for this study were considered to be in areas where the water does not appear to have been influenced by wastewater disposal. Because gross alpha- and beta-particle activity are less than the reporting levels for most of the samples, no statistical trend analyses were done. A few of the samples had concentrations equal to or greater than the reporting level. Some of the samples with results at reporting levels of 3s could be due to statistical fluctuations because the counting data are close to background concentrations. Other concentrations may represent background concentrations in the aquifer that have been detected because of increased sensitivity of the analyses, and some may be false positives (Davis, 2010, p. 28).

Calcium.—Calcium is a common cation in water that results from dissolution of minerals in the rock material that the water has contacted. Samples are not routinely analyzed for calcium at the INL because calcium is not considered a major by-product of wastewater disposal practices; however, calcium was collected routinely from three surface-water sites from 1965 to 1985.

Plots of calcium concentrations for the three surface-water sites (Big Lost River near Arco, Little Lost River near Howe, and Mud Lake near Terreton, fig. 1) are shown in appendix C. Summary statistics and trends calculated are given in table 3. No significant trend is observable from the data.

Magnesium.—Magnesium is a common cation in water that results from dissolution of minerals in the rock material that the water has contacted. Samples are not routinely analyzed for magnesium at the INL because it is not considered a by-product of wastewater disposal practices; however, magnesium was collected routinely from three surface-water sites from 1965 to 1985.

Plots of magnesium concentrations for the three surface-water sites (Big Lost River near Arco, Little Lost River near Howe, and Mud Lake near Terreton, fig. 1) are shown in appendix C. Summary statistics and trends calculated are given in table 3. No significant trend is observable from the data.

Potassium.—Potassium is a common cation in water that results from dissolution of minerals in the rock material that the water has contacted. Samples are not routinely analyzed for potassium at the INL because it is not considered a by-product of wastewater disposal practices; however, potassium was collected routinely from three surface-water sites from 1965 to 1985.

Plots of potassium concentrations for the three surface-water sites (Big Lost River near Arco, Little Lost River near Howe, and Mud Lake near Terreton, fig. 1) are shown in appendix C. Summary statistics and trends calculated are given in table 3. No significant trend is observable from the data.

Silica.—Dissolved silica is present in water from dissolution of minerals in the rock material that the water has contacted. Samples are not routinely analyzed for silica at the INL because it is not considered a by-product of wastewater disposal practices; however, silica was collected routinely from three surface-water sites from 1965 to 1985.

Plots of silica concentrations for the three surface-water sites (Big Lost River near Arco, Little Lost River near Howe, and Mud Lake near Terreton, fig. 1) are shown in appendix C. Summary statistics and trends calculated are given in table 3. No significant trend is observable from the data.

Sodium.—Sodium has been discharged in wastewater at INL facilities since they were established. The background concentration of sodium in water from the ESRP aquifer near the INL generally is less than 10 mg/L (Robertson and others, 1974, p. 155). Samples were routinely collected and analyzed for sodium from 66 aquifer wells and 3 surface-water sites used in this study. Summary statistics and trends calculated are given in table 3.

Plots of sodium concentrations for 66 wells and 3 surface-water sites are shown in appendix C. Sodium analyses were performed by various laboratories until 1989, and most analyses since have been done at the NWQL. The datasets include some outlier values, and if the value did not seem reasonable compared with the specific conductance, the outlier value was not used in the trend analyses. Sodium trends in water from sites at the INL are shown in figure 5. Results indicate that most of the sites show no statistical trend. A few of the wells influenced mostly by regional groundwater indicate an increasing trend that may indicate an influence from upgradient agricultural activities. A few wells (USGS 5, 12, and 15, appendix C) show variable increasing and decreasing patterns that probably are due to above-average and below-average periods of recharge to the aquifer. Analyses of trends for sodium in water from several of the wells near the NRF, ATRC, and INTEC indicate increasing trends, and although other data from these wells do not seem to show a direct influence of wastewater disposal practices, the increases possibly are due to long-term disposal practices.

Bromide.—Bromide has not been a major constituent discharged in wastewater at the INL; however, about 500 kg of bromine was discharged in liquid effluent at the INL from 1971 to 1998 (French and others, 1999). Bromide was routinely analyzed for water samples collected from eight wells on a quarterly basis as part of a NRF groundwater study from 1989 to 1995. Summary statistics and trends calculated are given in table 3.

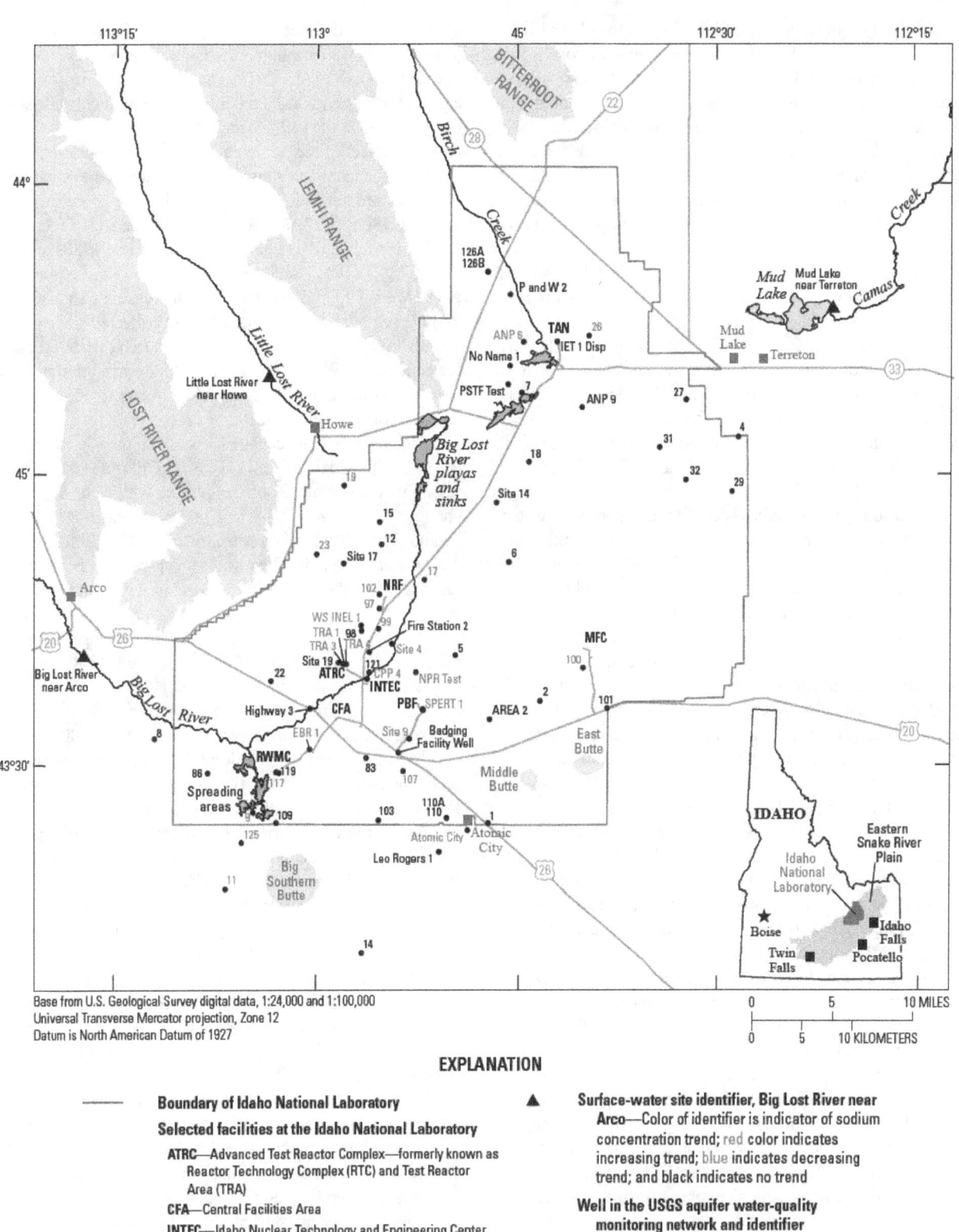

Base from U.S. Geological Survey digital data, 1:24,000 and 1:100,000
Universal Transverse Mercator projection, Zone 12
Datum is North American Datum of 1927

EXPLANATION

——— **Boundary of Idaho National Laboratory**

Selected facilities at the Idaho National Laboratory

ATRC—Advanced Test Reactor Complex—formerly known as
   Reactor Technology Complex (RTC) and Test Reactor
   Area (TRA)

CFA—Central Facilities Area

INTEC—Idaho Nuclear Technology and Engineering Center

MFC—Materials and Fuels Complex

NRF—Naval Reactors Facility

PBF—Power Burst Facility

RWMC—Radioactive Waste Management Complex

TAN—Test Area North

▲ **Surface-water site identifier, Big Lost River near
   Arco**—Color of identifier is indicator of sodium
   concentration trend; red color indicates
   increasing trend; blue indicates decreasing
   trend; and black indicates no trend

**Well in the USGS aquifer water-quality
   monitoring network and identifier**

   ANP 9 ●
      14

Local well identifier—Number only shows
   USGS well; other names are local well names.
   See table 2 for more information

Figure 5. Areal distribution of sodium concentration trends in water from selected wells and surface-water sites at and near
the Idaho National Laboratory, Idaho.

Plots of bromide concentrations for eight wells are shown in appendix C. Results for all but one site show an increasing trend in bromide concentration (table 3, appendix C). Hem (1989) indicates bromide is present at high concentrations in natural waters from seawater, geothermal sources, and from anthropogenic inputs, such as gasoline additives, fumigants, and fire-retardant agents. The increasing trends could be due to anthropogenic influences from the upgradient Little Lost River basin, or could be an indication of the influence of geothermal water in the NRF area. Analyses of trends for bromide in water from well WS INEL 1 indicate a decreasing trend, which is consistent with other constituents (chloride, sulfate, sodium, and nitrate) sampled from this well (table 3). A radioactive source was lost in the well after completion and initial sampling in 1978. The source had to be cemented in the bottom of the well and the decreasing trends possibly are not representative of the aquifer conditions, but are a remnant of contamination of the well from cementing in the source or from drilling fluids used to drill the nearby well INEL 1 (fig. 2).

Chloride.—Chloride has been discharged in wastewater at INL facilities since they were established. The background chloride concentration in water from the ESRP aquifer at the INL generally is about 15 mg/L (Robertson and others, 1974, p. 150). The secondary maximum contaminant level (MCL) for chloride in drinking water is 250 mg/L (U.S. Environmental Protection Agency, 2001). Samples were routinely collected and analyzed for chloride from all 67 aquifer wells and 7 surface-water sites used in this study. Summary statistics and trends calculated are given in table 3.

Plots of chloride concentrations for all sites used in this study are shown in appendix C. Chloride analyses were performed by various laboratories until 1989, and most analyses since 1989 have been done at the NWQL. The datasets include some outlier values, and if the value did not seem reasonable compared with the specific conductance of the well, the outlier value was not used in the trend analyses. Chloride trends in water from sites at the INL are shown in figure 6. Results indicate that water influenced by Big Lost River seepage has decreasing trends or, in some cases, has trends that correlate with above-average and below-average periods of recharge. Water samples from four sites on the Big Lost River also show a decreasing trend, or no trend, and concentrations generally are much less (mean concentrations of 3–6 mg/L) than those in the aquifer (about 15 mg/L). As large fluxes of Big Lost River seepage mix with the groundwater, the overall concentration in the groundwater near the river also likely will decrease if the well is obtaining water from the upper part of the aquifer. Analyses of trends for chloride in water from several of the wells that is mostly regionally derived groundwater generally indicate an increasing trend. These increasing trends are attributed to agricultural or other anthropogenic influences upgradient of the INL. Well USGS 4 shows a decreasing trend and the decrease probably is because the well was used for

a sodium chloride tracer study after originally being drilled in 1950 (Jack Barraclough, U.S. Geological Survey, retired, oral commun., May 25, 2011); therefore, the decreasing concentration trends are a result of the sodium chloride still being diluted in the well by regional groundwater.

The effect of the particular laboratory and/or the method used to collect samples is evident when evaluating concentration trends in a few wells. Chloride concentrations in wells USGS 6, 27, and 101 for the full period of record were compared to data collected since the late 1980s and early 1990s (fig. 7), when dedicated pumps were installed in the wells. Wells USGS 6 and 101 show increasing trends similar to other regional wells when only data collected since pump installation are considered, but show no trend when all data are considered. Well USGS 27 shows a decreasing trend when only more recent data are considered, but no trend when all data are considered.

The influence of seepage losses from the Big Lost River is especially evident in well USGS 12. Chloride concentrations (along with sodium and sulfate) plotted with water-level changes are shown in figure 8. Concentrations of chloride, sodium, and sulfate in water from well USGS 12 decrease during wetter periods when water levels rise, probably due to recharge from the river, and increase during drought periods when no recharge from the Big Lost River occurs. Figure 8 also shows how above-average and below-average periods of recharge (as indicated by flow in the Big Lost River) may affect concentration trends. Well USGS 32 is located in the northeastern part of the INL, far from any influence from the Big Lost River. Busenberg and others (2001, p. 84) indicated that the young fraction of water in this well was recent recharge from infiltration water, so periodic increases and decreases in this well are more likely influenced by the amount of infiltration, with above-average infiltration periods corresponding to decreases in constituent concentrations. Busenberg and others (2001) calculated relative ages of the young fraction of water for most of the wells sampled for this study and a more detailed comparison with ages are discussed in the section, "Significance of Trends at Each Well or Site."

Fluoride. —Fluoride in wastewater was discharged to percolation ponds at the INTEC during 1971–98 (Bartholomay and others, 2000). Background concentrations of fluoride in the ESRP aquifer in the southwestern part of the INL range from about 0.1 to 0.3 mg/L (Robertson and others, 1974, p. 75). The MCL for fluoride in drinking water is 4 mg/L (U.S. Environmental Protection Agency, 2001). Samples were routinely collected and analyzed for fluoride from nine wells and three surface-water sites used in this study. Summary statistics and trends calculated are given in tables 3 and 4.

Plots of fluoride concentrations for 12 sites are shown in appendixes C–E. Fluoride analyses were performed by various laboratories until 1989, and most analyses since then have been done at the NWQL. Results for all but one sample site (well USGS 15) show no trend.

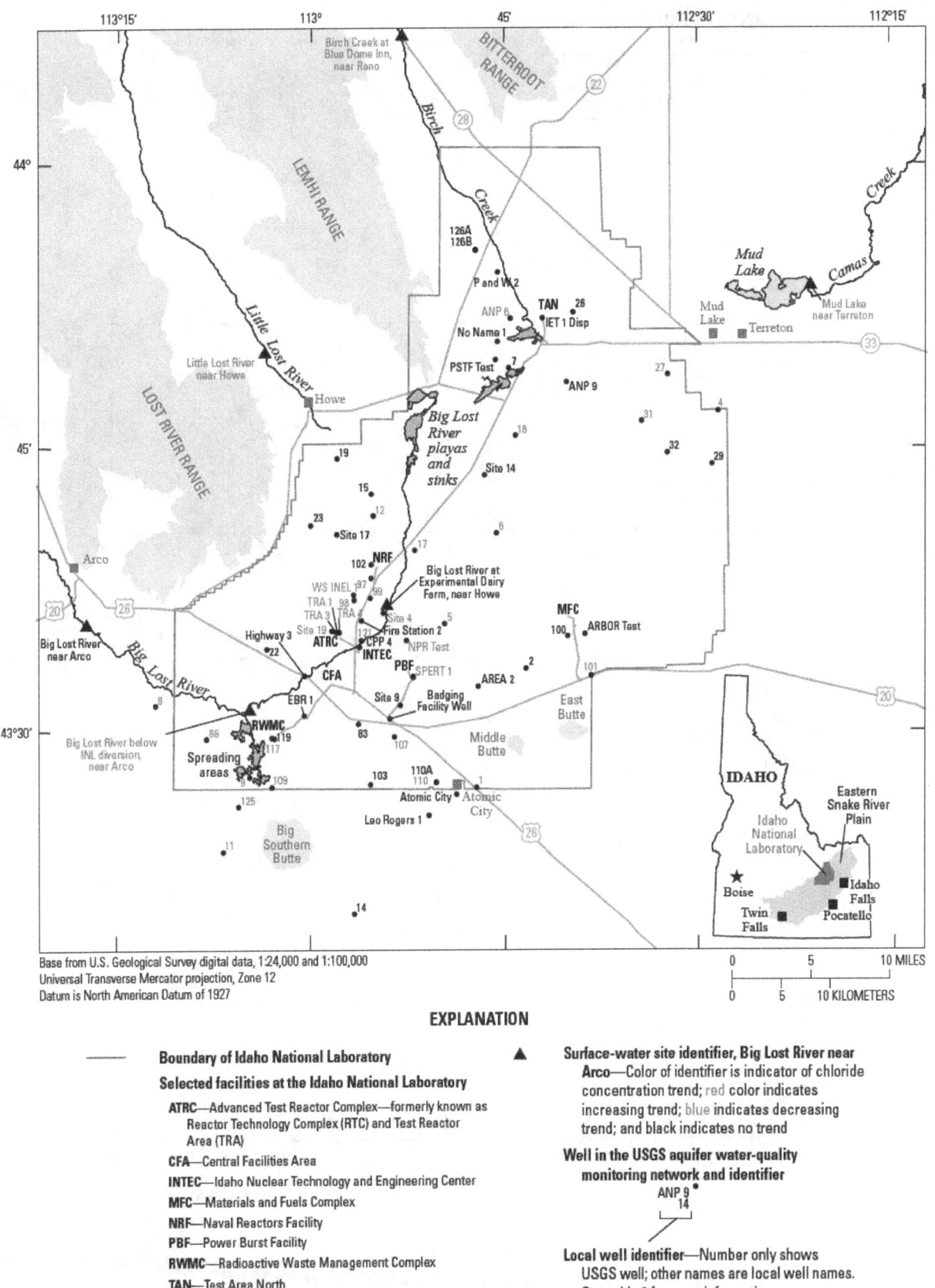

Figure 6.   Areal distribution of chloride concentration trends in water from selected wells and surface-water sites at and near the Idaho National Laboratory, Idaho.

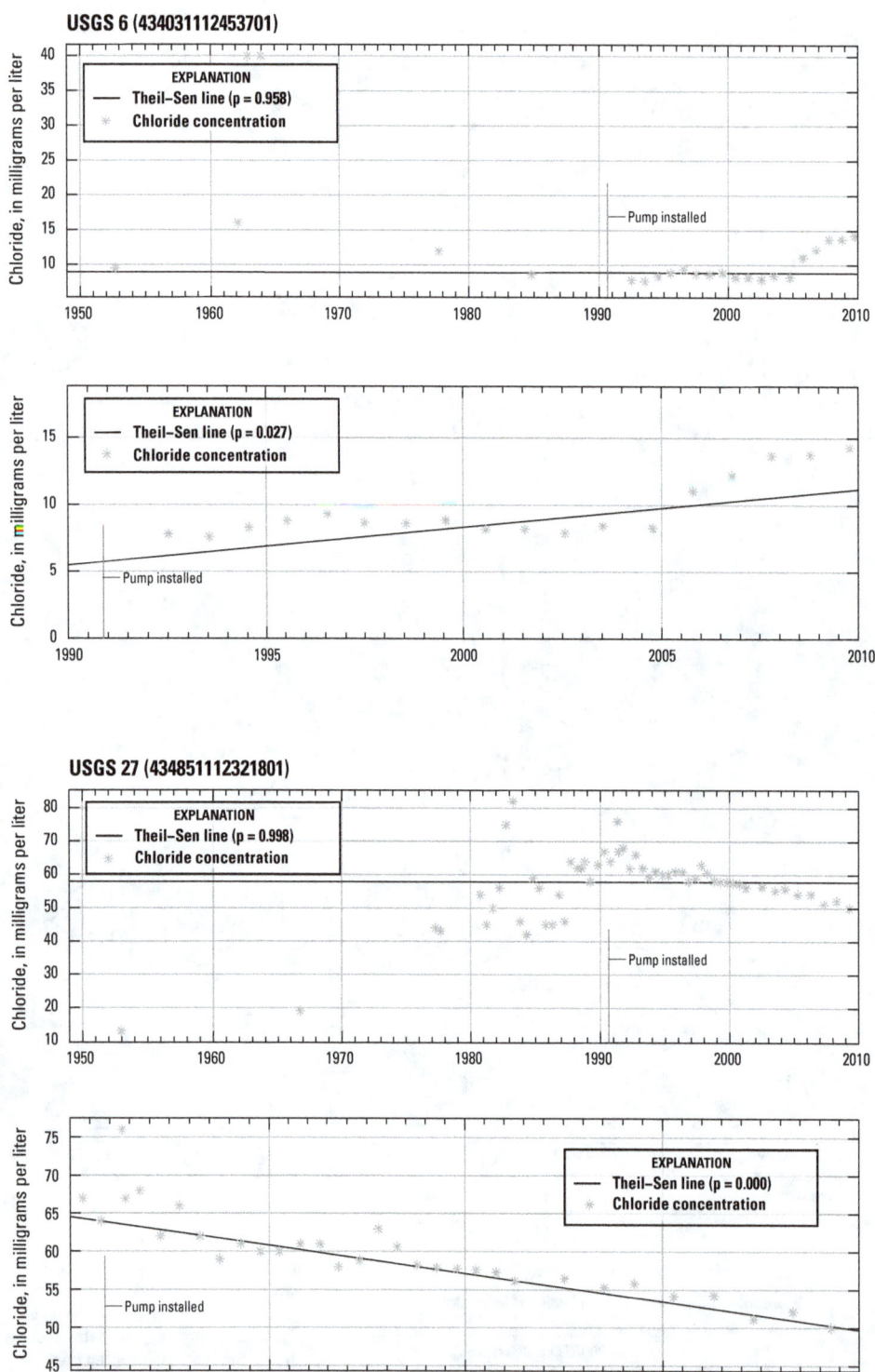

Figure 7. Chloride trends using the entire dataset versus data collected since pumps were installed in wells USGS 6, 27, and 101 at the Idaho National Laboratory, Idaho.

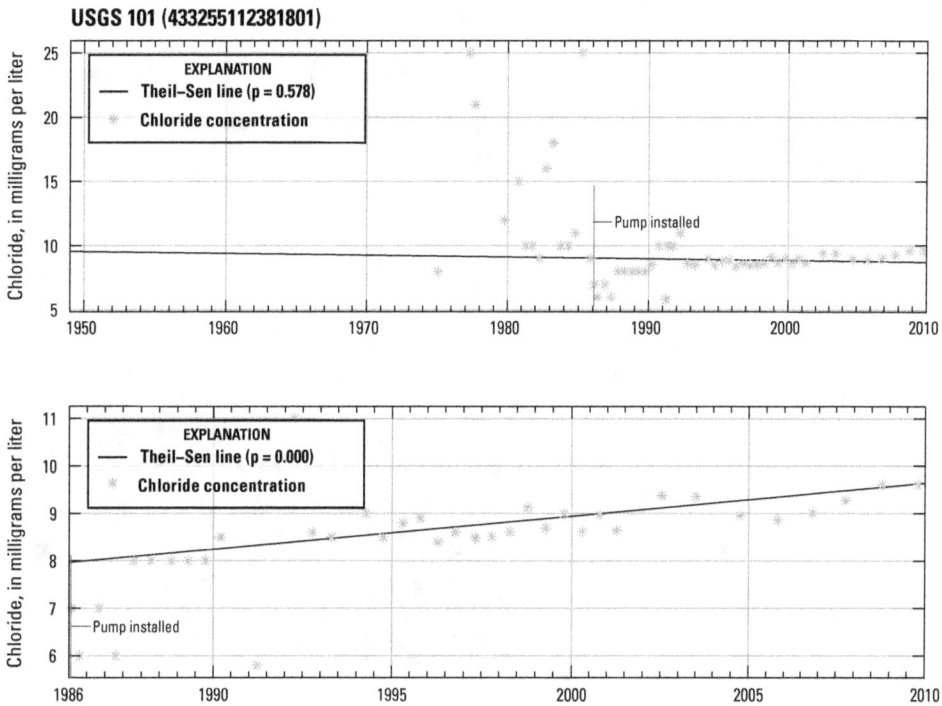

Figure 7.—Continued

Sulfate.—Sulfate has been discharged in wastewater at the INL facilities since they were established. The background sulfate concentrations in water from the ESRP aquifer in the south-central part of the INL range from about 10 to 40 mg/L (Robertson and others, 1974, p. 72). The secondary MCL for sulfate in drinking water is 250 mg/L (U.S. Environmental Protection Agency, 2001). Sulfate was routinely analyzed for water samples from 29 wells and 3 surface-water sites used in this study. Summary statistics and trends calculated are given in table 3.

Plots of sulfate concentrations for 32 sites are shown in appendix C. Sulfate analyses were performed by various laboratories until 1989, and most analyses since then have been done at the NWQL. The datasets include some outlier values, and if the value did not seem reasonable compared with the specific conductance of the well, the outlier value was not used in the trend analyses. Sulfate trends in water from sites at the INL are shown in figure 9.

Results indicate that wells influenced by seepage from the Big Lost River show variable responses, with some wells showing increases, some wells showing decreases, some wells showing influence from above-average and below-average periods of recharge, and some wells showing no trend. The Big Lost River water samples from one location (Big Lost River near Arco) did not show a trend. Analyses of trends

for sulfate in water from several of the wells in the eastern part of the INL that is mostly regionally derived groundwater indicate an increasing trend. These increased concentrations are attributed to agricultural or other anthropogenic influences upgradient of the INL.

Nitrate Plus Nitrite (as N).—Wastewater containing nitrate was injected into the ESRP aquifer through the INTEC disposal well from 1952 to February 1984, and was discharged to the INTEC percolation ponds after February 1984 (Orr and Cecil, 1991). Because nitrite analyses indicate that almost all nitrite plus nitrate concentrations are nitrate (as N), concentrations will be referred to as nitrate in this report. Concentrations of nitrate in groundwater not affected by wastewater disposal from INL facilities generally are less than 1 mg/L (as N) (Robertson and others, 1974, p. 73). The MCL for nitrate in drinking water is 10 mg/L (U.S. Environmental Protection Agency, 2001). Nitrate (as N) was routinely analyzed for water samples from 55 wells and 3 surface-water sites used in this study. Plots of nitrate concentrations for 58 sites are shown in appendix C. Summary statistics and trends calculated are given in table 3. Nitrate analyses were performed by various laboratories until 1989, and most analyses since then have been done at the NWQL. Nitrate trends in water from sites at the INL are shown in figure 10.

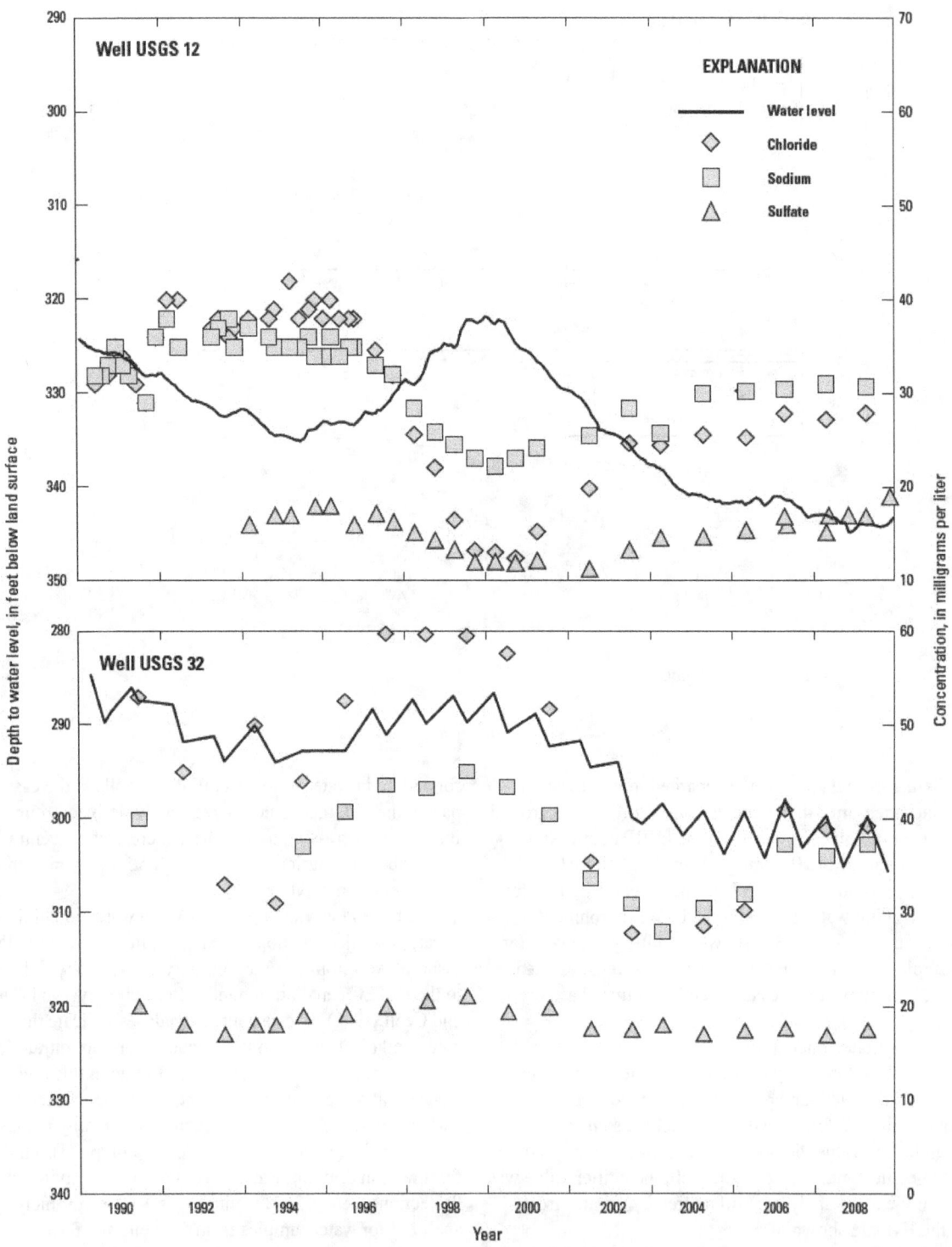

Figure 8.   Variable increases and decreases of chloride, sodium, and sulfate concentrations relative to water-level changes at wells USGS 12 and 32 and changes in flow from the Big Lost River, Idaho.

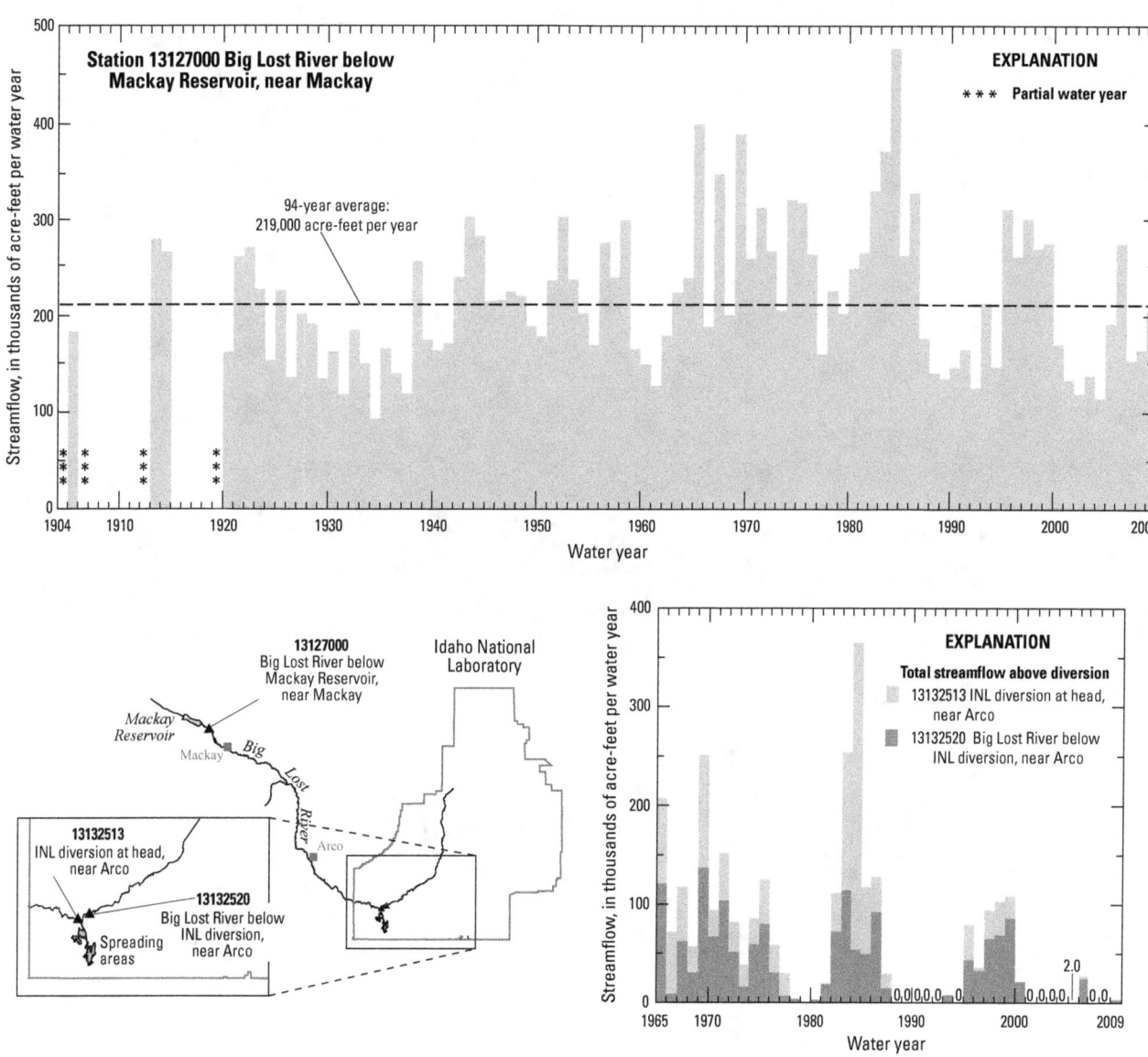

Figure 8.—Continued

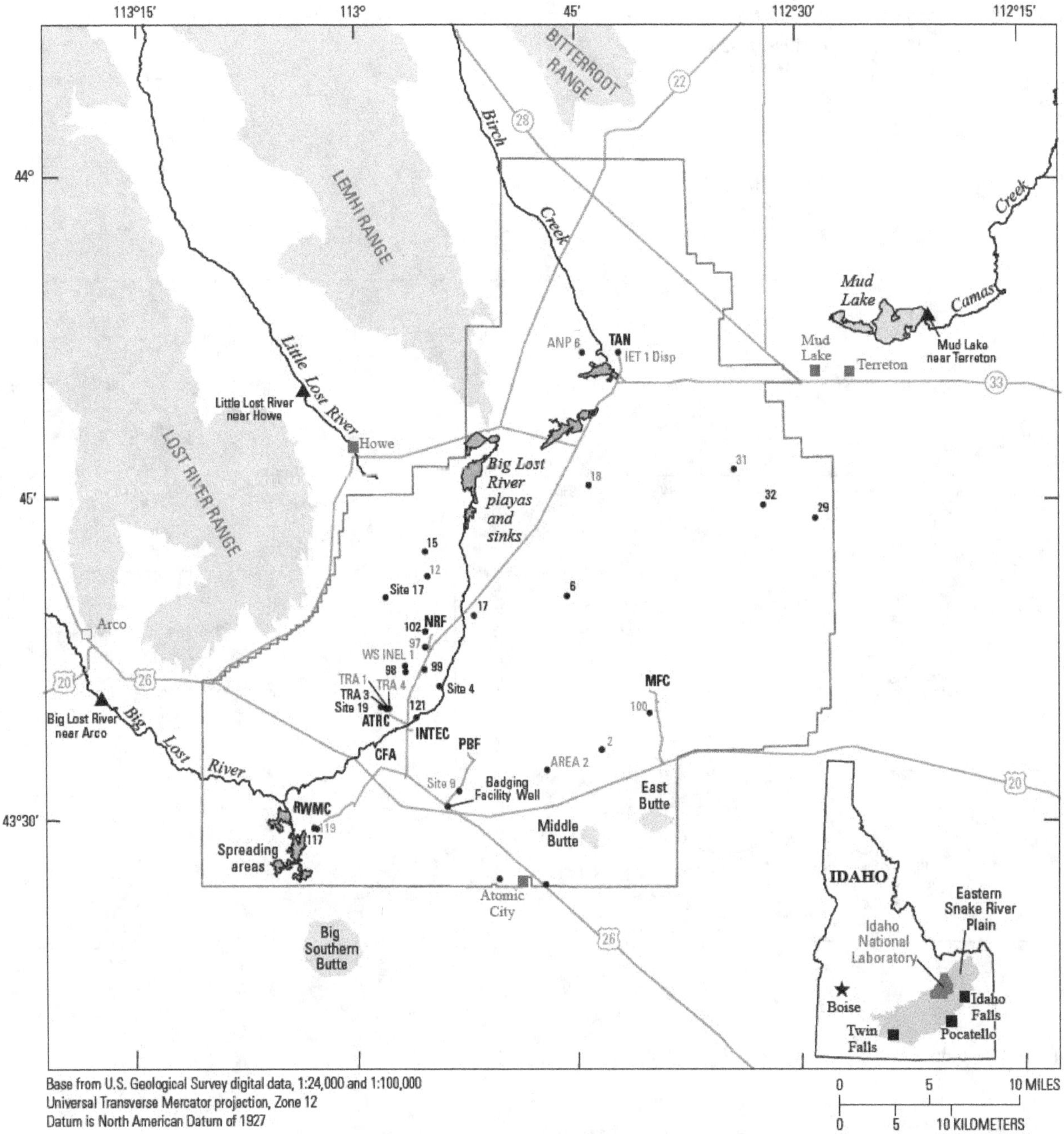

EXPLANATION

————— **Boundary of Idaho National Laboratory**

Selected facilities at the Idaho National Laboratory

    ATRC—Advanced Test Reactor Complex—formerly known as
    Reactor Technology Complex (RTC) and Test Reactor
    Area (TRA)

    CFA—Central Facilities Area

    INTEC—Idaho Nuclear Technology and Engineering Center

    MFC—Materials and Fuels Complex

    NRF—Naval Reactors Facility

    PBF—Power Burst Facility

    RWMC—Radioactive Waste Management Complex

    TAN—Test Area North

▲   **Surface-water site identifier, Big Lost River near
    Arco**—Color of identifier is indicator of sulfate
    concentration trend; red color indicates
    increasing trend; blue indicates decreasing
    trend; and black indicates no trend

**Well in the USGS aquifer water-quality
monitoring network and identifier**
    ANP 9
    14

Local well identifier—Number only shows
    USGS well; other names are local well names.
    See table 2 for more information

Base from U.S. Geological Survey digital data, 1:24,000 and 1:100,000
Universal Transverse Mercator projection, Zone 12
Datum is North American Datum of 1927

Figure 9.   Areal distribution of sulfate concentration trends in water from selected wells and surface-water sites at and near
the Idaho National Laboratory, Idaho.

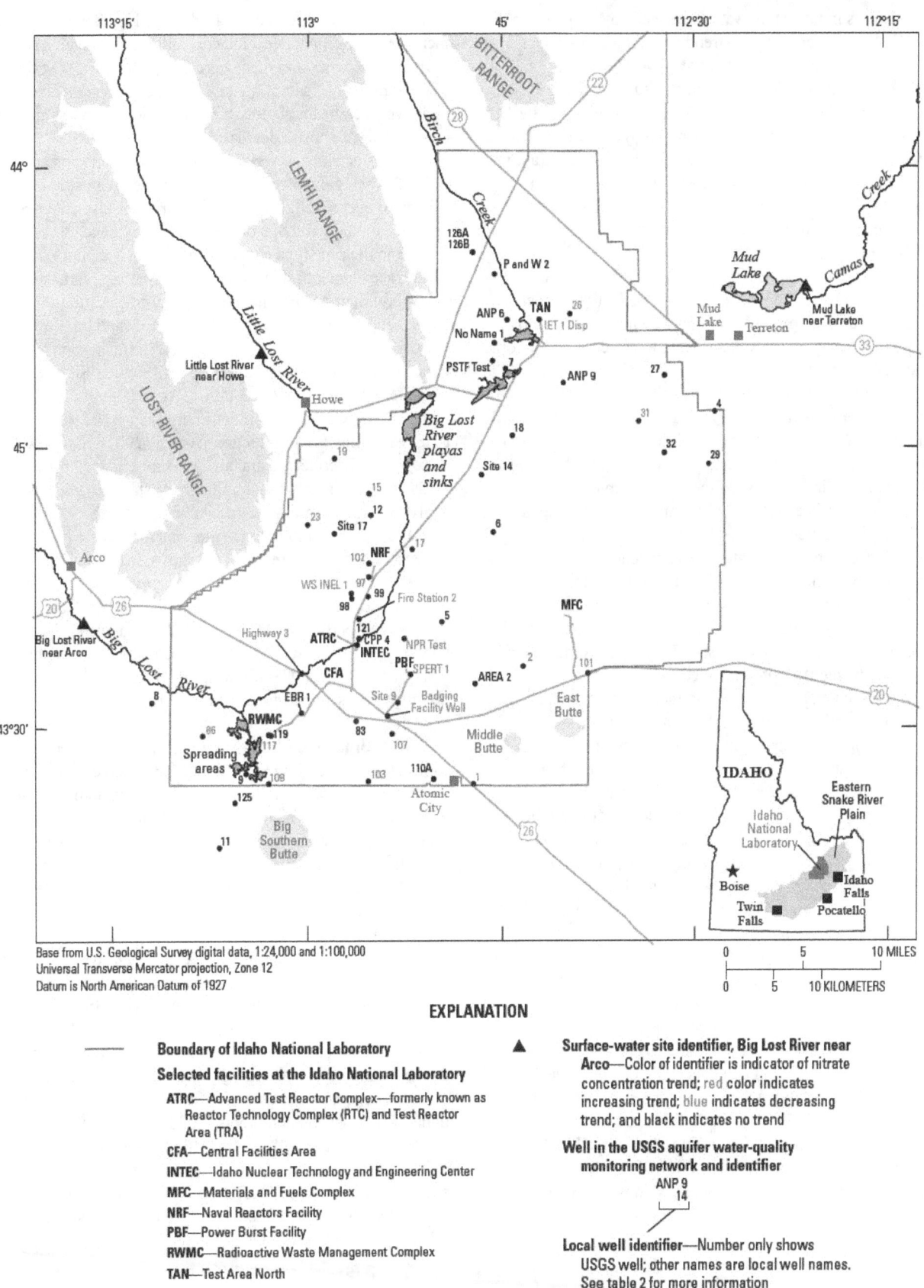

Figure 10. Areal distribution of nitrate concentration trends in water from selected wells and surface-water sites at and near the Idaho National Laboratory, Idaho.

Results indicate that water influenced by seepage from the Big Lost River shows no trend or, in some cases, has trends that may correlate with above-average flow or no flow in the Big Lost River and the spreading areas south of the RWMC. The water samples from one location on the Big Lost River (Big Lost River near Arco) did not show a trend; however, the Big Lost River has mean concentrations of nitrate (0.283 mg/L for Big Lost River near Arco, table 3) that are much less than aquifer wells near the Big Lost River. Nitrate concentrations appear to fluctuate periodically in some wells, which may be due to above-average and below-average recharge to the groundwater (appendix C). Analyses of trends for nitrate in water from several of the wells that is mostly regionally derived groundwater indicate increasing trends. These increasing trends are attributed to agricultural and other anthropogenic influences upgradient of the INL.

Orthophosphate (as P).—Orthophosphate (as P) has not been a major constituent discharged in wastewater at the INL; only about 9 kg was discharged in liquid effluent at the INL from 1971 to 1998 (French and others, 1999). Orthophosphate (as P) was routinely analyzed for water samples from 55 wells and 3 surface-water sites. Summary statistics and trends calculated are given in tables 3 and 4.

Plots of orthophosphate (as P) concentrations for the 58 sites are given in appendixes C–E. Because orthophosphate (as P) concentrations were equal to or less than the reporting levels for many of the samples, trend analyses were done for censored and uncensored datasets. Trend results indicate no trend for most of the sites sampled.

Trace Elements.—A suite of trace elements including aluminum, antimony, arsenic, barium, beryllium, cadmium, cobalt, copper, lead, manganese, molybdenum, nickel, silver, uranium, and zinc were routinely collected from seven wells analyzed for this study. Chromium was analyzed for water samples from these 7 wells along with 37 other wells so general discussion will be included in the section, "Chromium." Thallium was collected from five wells and selenium was collected from two wells. These trace elements are sampled for upgradient and downgradient wells at TAN and ATRC because they were considered constituents of concern in the 1993 INL groundwater-monitoring plan at those facilities (Sehlke and Bickford, 1993).

Plots of concentrations of trace elements from wells with both uncensored and censored data are shown in appendixes D and E. Summary statistics and trends calculated are given in tables 3 and 4. Lead concentrations decreased in well USGS 98. Molybdenum concentrations decreased in three wells near TAN (ANP 9, PSTF Test, and USGS 26). Molybdenum showed an increasing trend in well USGS 97. Barium concentrations decreased in wells USGS 7 and USGS 97. Selenium concentrations increased in well USGS 98.

Uranium concentrations decreased in the No Name 1 well and increased in well ANP 9. Zinc concentrations in well ANP 9 decrease, increase, and then decrease again throughout the 15-year history of sampling, and the fluctuations may be due to above-average and below-average periods of recharge. Zinc concentrations show a decreasing trend in well USGS 26. Zinc concentrations in well USGS 98 decreased substantially in 2005 when galvanized pipe was replaced with stainless steel pipe in the well. Barium concentrations also showed a decrease in 2005 and this decrease may be due to either the change in the pipe type or the deepening of the pump. All other trace elements showed no trends (tables 3 and 4).

Chromium.—Documented disposal of chromium in wastewater has occurred at ATRC, INTEC, and the Power Burst Facility (PBF) (Davis, 2010, p. 37). Background concentrations of chromium in water from the ESRP aquifer at the INL range from 2 to 3 µg/L (Orr and others, 1991, p. 41). The MCL for chromium in drinking water is 100 µg/L (U.S. Environmental Protection Agency, 2001). Samples were routinely collected and analyzed for chromium from 44 wells analyzed for this study. Summary statistics and trends analyzed for chromium are shown in tables 3 and 4.

Plots of chromium concentrations for wells with both uncensored and censored data are shown in appendixes C–E. Five wells located in the northern part of the INL near TAN (ANP 9, No Name 1, PSTF test, USGS 7, and USGS 26) show decreasing trends. Well WS INEL 1 shows a decreasing trend, which is consistent with most other constituents in water from that well. Well USGS 98 shows an increasing trend and all other sites sampled for chromium show no trend (tables 3 and 4).

Total Organic Carbon.—Water samples analyzed for total organic carbon (TOC) are used to screen for organic compounds in the aquifer as a general indicator of groundwater contamination. As part of the INL groundwater-monitoring program adopted in 1994, the USGS began collecting and analyzing water from several wells at the INL for TOC. TOC was routinely analyzed for water samples from 37 wells used in this study. Summary statistics for TOC results are shown in tables 3 and 4.

Plots of concentrations of TOC for water from wells with both uncensored and censored data are given in appendixes C and D. Most of the wells near the NRF (USGS 12, 17, 97, 98, 99, and 102) show increasing trends, and well USGS 19 at the terminus of the Little Lost River basin also shows an increasing trend. Increasing trends may be an indication of anthropogenic compounds in the Little Lost River basin underflow. Swanson and others (2002) were unable to model water from several wells in the Little Lost River basin because of contamination from local agricultural practices. Busenberg and others (2001, p. 86) indicated that the water chemistry in well USGS 12 was affected by agricultural chemicals. All other sites sampled for TOC show no trend (tables 3 and 4).

Significance of Trends at Each Well or Site

Chloride, nitrate, sodium, and sulfate are the primary constituents that have been discharged in wastewater at INL facilities and their concentrations in water samples often can be affected by anthropogenic influences. Because of this, these are the primary constituents discussed for each well or site.

ANP 6.—Analyses of trends for constituents in water from this well indicate that chloride and nitrate concentrations generally were increasing until the late 1990s and early 2000s. Since then, the concentrations have decreased (appendix C). The gradual increases and decreases in concentrations in this well seem to indicate that the well is influenced by above-average and below-average periods of recharge. As discussed previously, concentrations of constituents probably decrease during wet periods, when the aquifer is diluted more with low-concentration infiltration water, and may increase during drought periods when the aquifer system is influenced by older water that has had more time to react with chemicals in aquifer material. Busenberg and others (2001, p. 141) indicated that the age of the young fraction of water in this well during 1995–96 probably was between 20 and 30 years old, so low concentrations in this well in the early 1990s and late 2000s likely correspond to the wet periods in the mid-1970s and mid-1980s, thus, future sampling should continue for these constituents in water from this well to further test this theory.

ANP 9.—Analyses of trends for constituents in water from this well indicate no trends for chloride, sodium, and nitrate concentrations (table 3). Busenberg and others (2001, p. 81) indicated that water in this well is virtually all old regional water. Because of the lack of significant long-term temporal trends, a reduced sampling frequency may be considered.

ARBOR Test.—Analyses of trends for constituents in water from this well indicate no trend for chloride concentrations (table 3). Busenberg and others (2001, p. 80) indicated that water in this well is a mixture of old regional water and young infiltration water. Given the distance of the well from any streamflow infiltration and the probable small amount of infiltration recharge even during wet periods, a reduced sampling frequency may be considered.

AREA 2.—Analyses of trends for constituents in water from this well indicate no trends for sodium, chloride, and nitrate concentrations (table 3). Busenberg and others (2001, p. 80) indicated that water in this well is a mixture of old regional and young infiltration water. Given the distance of the well from streamflow infiltration and the relatively small amount of infiltration recharge even during wet periods, a reduced sampling frequency may be considered.. The increasing trend of sulfate is similar to overall increases of this constituent in the regional aquifer system, and may be the result of agricultural influences on the regional aquifer system.

Atomic City.—Analyses of trends for constituents in water from this well indicate no trend for chloride concentrations, and an increasing trend for sodium concentrations (table 3). Busenberg and others (2001, p. 80) indicated that water in this well is a mixture of old regional water somewhat influenced by young infiltration water. Given the distance of the well from streamflow infiltration and the probable small amount of infiltration recharge even during wet periods, a reduced sampling frequency may be considered. It is uncertain why sodium concentrations may be increasing.

Badging Facility Well.—Analyses of constituent trends in water from this well indicate no trends for chloride, sodium, and sulfate concentrations, and a decreasing trend for nitrate concentrations (table 3); however, some minor varibility in concentrations has occurred during different timeframes (appendix C). Busenberg and others (2001, p. 84) indicated that CFC concentrations may be influenced by waste disposal and, given the well location, the small variability in constituent concentrations could be due to different disposal rates at upgradient facilities.

CPP 4.—Analyses of constituent trends in water from this well indicate no trends for chloride and nitrate concentrations, and an increasing trend for sodium concentrations (table 3). This well is located near the Big Lost River; however, above-average and no-flow periods seem to have little influence on the constituents in the water, with the possible exception of chloride (appendix C). The lack of significant trends may be because this well is open to a large section of the aquifer that combines a mixture of water types.

EBR 1.—Analyses of constituent trends in water from this well indicate no trends for chloride and nitrate concentrations, and an increasing trend for sodium concentrations (table 3). The lack of significant trends for most constituents may be because this well is open to a large section of the aquifer that combines a mixture of water types. Because this well is downgradient of the INTEC and the ATRC, the increasing sodium trend possibly could be due to wastewater disposal; however, no other constituents support this concept and sodium concentrations in the well are less than aquifer background concentrations.

Fire Station 2.—Analyses of constituents in water from this well indicate no trend for chloride, although there has been some variability in concentrations over time, (appendix C), an increasing trend for nitrate concentrations, and no trend for sodium concentrations (table 3). The variability of chloride concentrations could be due to fluctuations related to above-average and below-average periods of recharge or changes in the amount of wastewater disposal at the NRF. Busenberg and others (2001, p. 141) indicated that the young fraction of water in this well probably is between 11 and 20 years old, so the low chloride concentrations in the late 1980s would correspond to above-average periods of recharge in the mid-1970s. Nitrate concentrations greater than 1 mg/L indicate some anthropogenic influence at this well, and, given its location, the increasing trend could be due to wastewater disposal at the NRF. Because power to the pump house was disconnected in 1997, the USGS discontinued sampling at this well at that time.

Highway 3.—Analyses of constituent trends in water from this well indicate no trends for chloride and sodium concentrations, and an increasing trend for nitrate concentrations (table 3). This well is located near the Big Lost River; however, above-average and no-flow periods seem to have little influence other than some minor changes in sodium and chloride concentrations (appendix C). The lack of significant trends may be because this well is open to a large section of the aquifer that combines a mixture of water types.

IET 1 Disp.—Analyses of trends for constituents in water from this well indicate no trends for chloride and sodium concentrations, and an increasing trend for sulfate concentrations (table 3). The nutrient concentrations (nitrate, orthophosphate, and ammonia) in this well show decreasing trends. This well was completed with a slotted screen that has corroded and allowed sediment from above the water table to fill in the bottom of the well; thus, samples will not be able to be collected in the future unless well rehabilitation is done or water levels rise. Because of these factors, sampling has been discontinued.

Leo Rogers 1.—Analyses of trends for constituents in water from this well indicate no trends for sodium and chloride concentrations. Busenberg and others (2001, p. 80) indicated that water in this well is a mixture of old regional water influenced by young infiltration water. The well initially was used as an irrigation well and is open to a large section of the aquifer. Sampling of this well has been discontinued because of the lack of trends and because power is no longer available.

No Name 1.—Analyses of trends for constituents in water from this well indicate no trends for chloride, nitrate, and sodium concentrations (table 3). Busenberg and others (2001, p. 83) indicated that water in this well is mostly old slow infiltration recharge water that is different from many other water types in the northern area of the INL. Because of the lack of significant long-term temporal trends, a reduced sampling frequency may be considered.

NPR Test.—Analyses of trends for constituents in water from this well indicate decreasing trends for chloride, sodium, and nitrate concentrations. Geochemical modeling at this well (Schramke and others, 1996) indicates that the water from this well is partially seepage loss from the Big Lost River, and decreasing trends may be attributed to the dilution of the water from the lower concentration river water. Chloride samples collected prior to 1990 may show an increase until 1991 (appendix C), prior to the overall decreasing trend. Busenberg and others (2001, p. 142) indicated that the age of the young fraction of water in this well was about 25 years and, given this age, the peak concentrations in 1991 would correspond with drought conditions in the mid-1960s.

P and W 2.—Analyses of trends for constituents in water from this well indicate no trends for chloride, sodium, and nitrate concentrations (table 3). Busenberg and others (2001, p. 82) indicated that this well consists of mostly one type of water. The temperature of this water generally is colder than water in other wells in the aquifer and the colder temperature probably can be attributed to Birch Creek recharge.

PSTF Test.—Analyses of trends for constituents in water from this well indicate no trends for chloride, nitrate, and sodium concentrations (table 3). Busenberg and others (2001, p. 82) indicated that water in this well is mostly old, slow infiltration recharge water that is different from water in the nearby well P and W 2 in the northern area. Because of the lack of significant long-term temporal trends, a reduced sampling frequency may be considered.

Site 4.—Analyses of trends for constituents in water from this well indicate decreasing trends for chloride and sodium, and no trend for sulfate; however, there has been considerable variability in the concentrations of all three constituents over time (appendix C). The variable decreases and increases in this well may indicate that the well is influenced by above-average and below-average periods of recharge. The well is located near the Big Lost River, and the young fraction of water is believed to be about 25 years old (Busenberg and others, 2001, p. 142).

Site 9.—Analyses of trends for constituents in water from this well indicate increasing trends for sodium, sulfate, and nitrate concentrations, and no trend for chloride concentrations (table 3). Busenberg and others (2001, p. 142) indicate that the age of the young fraction of water was greater than 35 years old, but probably a much younger age near the surface of the water table. Concentration increases could be due to anthropogenic influences on the regional aquifer system, or could be due to wastewater disposal in the PBF area.

Site 14.—Analyses of trends for constituents in water from this well indicate no trends for chloride, nitrate, and sodium concentrations (table 3). Busenberg and others (2001, p. 82) indicated that water in this well is mostly old water. The well is open to a large section of the aquifer, so a reduced sampling frequency may be considered.

Site 17.—Analyses of constituents in water from this well indicate no trends for concentrations of chloride, nitrate, sodium, and sulfate; however, there has been considerable variability in the concentrations of all four constituents over time (appendix C). The temporal variability in water-quality conditions in this well seem to indicate that the well is influenced by above-average and below-average periods of recharge, possibly from the Big Lost River. Busenberg and others (2001, p. 142) indicate that the young fraction of water is about 21–22 years old, so high concentrations in the 2000 timeframe probably correspond with the drought in the early 1980s.

Site 19.—Analyses of trends for constituents in water from this well indicate a decreasing trend for chloride concentrations, and no trends for sodium and sulfate concentrations (table 3). Concentrations of chloride, sodium, and sulfate increased for a couple of samples in the early 2000s, but no other indications of above-average or below-average periods of recharge are evident in this well (appendix C). Busenberg and others (2001, p. 85) indicated that water in this well probably contains no regional water, so water from this well probably is representative of mixed tributary valley underflow.

SPERT 1.—Analyses of trends for constituents in water from this well indicate increasing trends for chloride, sodium, and nitrate concentrations (table 3). Because some wastewater disposal has occurred throughout the history of all facilities in the SPERT 1 area, the increasing trends possibly are from infiltration of wastewater discharged in the past finally reaching the aquifer or from anthropogenic influences on the regional aquifer system.

TRA 1.—Analyses of trends for constituents in water from this well indicate a decreasing trend for chloride concentrations, and an increasing trend for sodium and sulfate concentrations (table 3). The decreasing chloride concentrations are consistent with other long-term decreasing trends associated with Big Lost River seepage in this part of the aquifer. The increasing trend for sodium and sulfate could be due to wastewater discharge at the ATRC.

TRA 3.—Analyses of trends for constituents in water from this well show a decreasing trend for chloride concentrations, an increasing trend for sodium concentrations, and no trend for sulfate concentrations (table 3). The decreasing chloride concentrations are consistent with other long-term decreasing trends associated with Big Lost River seepage in this part of the aquifer. The increasing trend for sodium could be due to wastewater discharge at the ATRC.

TRA 4.—Analyses of trends for constituents in water from this well indicate a decreasing trend for chloride concentrations, and increasing trends for sodium and sulfate concentrations (table 3). The decreasing chloride concentrations are consistent with other long-term decreasing trends associated with Big Lost River seepage in this part of the aquifer. The increasing trends for sodium and sulfate concentrations could be due to wastewater discharge at the ATRC.

USGS 1.—Analyses of trends for constituents in water from this well indicate increasing trends for chloride and nitrate concentrations, and no trend for sodium concentrations (table 3). Busenberg and others (2001, p. 80) indicated that water in this well is about 40 percent young recharge water and 60 percent regional recharge water, and the increasing trends for chloride and nitrate concentrations may be due to regional anthropogenic influence on the aquifer.

USGS 2.—Analyses of trends for constituents in water from this well indicate no trend for chloride and sodium concentrations, and increasing trends for sulfate and nitrate concentrations (table 3). Busenberg and others (2001, p. 80) indicated that water in this well is about 40–50 percent young recharge water, and the remainder is regional recharge water. The increasing trends for sulfate and nitrate concentrations may be due to regional anthropogenic influences on the aquifer.

USGS 4.—Analyses of trends for constituents in water from this well indicate a decreasing trend for chloride concentrations, and no trend for the other constituents (table 3). The decreasing trend for chloride probably is because the well was used for a sodium chloride tracer study after originally being drilled (Jack Barraclough, U.S. Geological Survey retired, oral commun., May 25, 2011); however, the reason for the lack of a similar trend for sodium concentrations is unknown.

USGS 5.—Analyses of trends for constituents in water from this well indicate an increasing trend for chloride; however, concentrations have been quite variable over time. (appendix C). There are no trends for sodium and nitrate concentrations (table 3). The variable changes for chloride probably are due to above-average and below-average periods of recharge. Busenberg and other (2001, p. 82 and 142) indicated that water in this well is about 27 percent regional recharge water, and the remainder is mostly younger (about 16 years old) recharge water from Big Lost River sinks. The increasing chloride concentrations in the late 2000s correlate well with drought conditions in the early 1990s.

USGS 6.—Analyses of trends for constituents in water from this well indicate an increasing trend for chloride concentrations, and no trends for nitrate, sodium, and sulfate concentrations (table 3). Busenberg and others (2001, p. 83) indicated that water in this well is all old water, so it is uncertain as to why the chloride concentrations are increasing.

USGS 7.—Analyses of trends for constituents in water from this well indicate no trends in concentrations for most of the constituents (table 3). Busenberg and others (2001, p. 83) indicated that water in this well is all old water from a large section of the aquifer, so a reduced sampling frequency may be considered.

USGS 8.—Analyses of trends for constituents in water from this well indicate a decreasing trend in chloride concentrations, and no trends for sodium and nitrate concentrations (table 3). Variability of sodium and chloride concentrations is evident in the datasets (appendix C). Busenberg and others (2001, p. 92) indicated that the young fraction of water in this well was about 8–9 years old and was related to Big Lost River streamflow, so it seems probable that some response from seepage when flow occurs in the river would be evident.

USGS 9.—Analyses of trends for constituents in water from this well indicate decreasing trends for chloride and sodium concentrations, and no trend for nitrate concentrations (table 3). Busenberg and others (2001, p. 86) indicated that most of the water from this well was younger water with a recharge age of 21–23 years, and about 20 percent was old regional water. They also speculated that the young recharge water was from the INTEC disposal well, so decreasing trends in chloride and sodium concentrations could be representative of decreased discharge of wastewater constituents at the INTEC from earlier time periods. This well is located in the INL Spreading Areas (fig. 2), but water levels do not show any major fluctuation when there is flow in the Spreading Areas, so decreases in chloride and sodium concentrations related to decreased wastewater disposal is more likely than decreases due to seepage from the Spreading Areas.

USGS 11.—Analyses of constituents in water from this well indicate decreasing trends for chloride and sodium concentrations, and no trend for nitrate concentrations (table 3). Busenberg and others (2001, p. 86) indicated that most of the water from this well was younger water with a recharge age of 17 years, and only a small fraction was regional groundwater. They also speculated that the young recharge water was from the INTEC disposal well, suggesting that decreasing chloride and sodium concentrations could be representative of decreased discharge of wastewater constituents at the INTEC.

USGS 12.—Analyses of trends for constituents in water from this well indicate decreasing trends for chloride and sulfate, and no trend for sodium or nitrate; however, concentrations for all four constituents have experienced considerable variability over time (appendix C). The gradual increases and decreases in concentrations in this well seem to indicate that the well is influenced by above-average seepage and no seepage from the Big Lost River, and this well shows large water-level fluctuations in response to seepage (fig. 8). Busenberg and others (2001, p. 86) indicated a discrepancy in the results from methods used to date the young fraction of water from this well, with tritium/helium ages of about 3–5 years and CFC ages of about 17 years. The rapid response of water-level changes when flow in the Big Lost River occurs seem to indicate the younger age may be more accurate as high and low concentrations correlate fairly well with low and high water levels, respectively (fig. 8). Schramke and others (1996) indicate that water in this well also was influenced geochemically by Little Lost River underflow, so above-average and below-average periods of recharge from the Little Lost River basin also may be a factor.

USGS 14.—Analyses of trends for constituents in water from this well show some variability but no trends for chloride and sodium concentrations (appendix C, table 3). Busenberg and others (2001, p. 86) indicated that most of the water from this well was local recharge with a recharge age of about 27 years. They also speculated that the young recharge water

was from the INTEC disposal well, so variability of chloride and sodium concentrations could be representative of variable discharge of wastewater containing these constituents at the INTEC.

USGS 15.—Analyses of constituents in water from this well indicate no trend for chloride, sodium, sulfate, and bromide and an increasing trend for nitrate concentrations; however, concentrations for all constituents have exhibited considerable variability over time (appendix C). The gradual increases and decreases in concentrations in this well seem to indicate that the well is influenced by above-average and below-average periods of seepage from the Big Lost River. However, Busenberg and others (2001, p. 86) indicated that water in this well is mostly old water, with only a small fraction of young water that is between 22 and 26 years old. The increases and decreases do not correlate well with Big Lost River seepage and the corresponding young water age. Schramke and others (1996) indicated that the nearby well USGS 12 (fig. 2) was influenced geochemically by Little Lost River underflow, so above-average and below-average periods of recharge from the Little Lost River basin may be a factor affecting the variability of the concentrations.

USGS 17.—Analyses of constituents in water from this well indicate increasing trends for sodium, nitrate, and TOC concentrations, no trend for sulfate concentrations, and a decreasing trend for chloride concentrations (tables 3 and 4). Busenberg and others (2001, p. 86) indicated that most of the water from this well was local recharge with a recharge age of about 16 years. The increasing trends for sodium, nitrate, and TOC concentrations could be due to anthropogenic influences from the Little Lost River basin.

USGS 18.—Analyses of trends for constituents in water from this well indicate increasing trends for chloride and sulfate concentrations, but no trends for other constituents (tables 3 and 4). Busenberg and others (2001, p. 83) indicated that most of the water from this well was old water. The increasing trends of chloride and sulfate concentrations probably are due to anthropogenic influences on the regional aquifer system northeast of the INL.

USGS 19.—Analyses of constituents in water from this well indicate no trend for chloride and decreasing trends for sodium and nitrate concentrations (table 3); however, all concentrations of all constituents have exhibited considerable variability over time (appendix C). Busenberg and others (2001) indicated that the young fraction of water was about 15–16 years old, but did not speculate on the source of water. Given the location of the well, the groundwater probably is influenced by Little Lost River seepage and underflow from the Little Lost River Valley. The variable changes for chloride could be due to above-average and below-average periods of recharge; however, since 1995 the concentrations have remained consistent. The decreasing sodium and nitrate concentrations could be an indication of greater influence of lower concentration surface-water seepage.

USGS 22.—Analyses of trends for constituents in water from this well indicate no trends in chloride and sodium concentrations (table 3). Busenberg and others (2001, p. 90) indicated that the water chemistry of this well is very different from other water at the INL, and probably is local recharge of both rapid-focused recharge and slow-infiltration recharge. Because of these recharge mechanisms, trends are not expected to change much in the future and sampling may be discontinued.

USGS 23.—Analyses of constituents in water from this well indicate no trend in chloride and nitrate concentrations and an increasing trend for sodium concentrations (table 3); however, there has been considerable variability in the concentrations of all three constituents over time. Busenberg and others (2001) indicated that water in this well is mostly old water and, given its location, probably is Little Lost River groundwater underflow. The variability implies changes due to above-average and below-average periods of recharge. Sodium increases may be due to anthropogenic influences from the Little Lost River valley.

USGS 26.—Analyses of trends for constituents in water from this well indicate increasing trends for sodium and nitrate concentrations, and no trend for chloride concentrations (table 3). Busenberg and others (2001, p. 83) indicated that water in this well is mostly all old water, so the increasing concentration trends probably are due to anthropogenic influences on regional recharge from northeast of the INL.

USGS 27.—Analyses of constituents in water from this well indicate a decreasing trend in chloride concentrations and no trends for nitrate and sodium concentrations (table 3); however, the concentrations of all three constituents have exhibited considerable variability over time. Busenberg and others (2001, p. 83) indicated that water in this well is mostly all old water with the young fraction about 10–13 years old. The general increase in chloride until about 1991, and then decrease since 1991 (appendix C, table 3) does not correlate well with above-average and below-average periods of recharge, so the variability could be due to pump installation in 1990.

USGS 29.—Analyses of trends for constituents in water from this well indicate no trends for all constituents (table 3); however, there appears to be some variability in nitrate concentrations (appendix C). Busenberg and others (2001, p. 83) indicated that the young fraction was about 28 years old and that samples showed a strong agricultural influence. The variability of nitrate concentrations may be due to variable agricultural influences.

USGS 31.—Analyses of trends for constituents in water from this well indicate increasing trends for chloride, nitrate, and sulfate concentrations, and no trend for sodium concentrations (table 3). Busenberg and others (2001, p. 84) indicated that water in this well is mostly old water with slow infiltration, so the increasing trends probably are due to anthropogenic influences on regional recharge from northeast of the INL.

USGS 32.—Analyses of constituents in water from this well indicate no trend in any constituent concentrations (table 3); however, concentrations for all four constituents have exhibited considerable variability over time. Concentrations tended to increase until about 2000, decrease until 2004, and then mostly increase again (appendix C). Busenberg and others (2001, p. 84) indicated that water in this well is mostly old water with slow infiltration, so the variable changes probably are due to agricultural influences on regional recharge from northeast of the INL and responses to above-average and below-average periods of recharge. This probably explains why higher concentrations of sodium, sulfate, and chloride in the 1997–2000 timeframe shown in figure 8 do not correlate with low water levels in the aquifer because the water sampled during that wet period probably was recharged during a previous dry period.

USGS 83.—Analyses of constituents in water from this well indicate no trends for any constituent (table 3); however, chloride and nitrate concentrations variably decreased until about 2000, increased until 2004, and then decreased again (appendix C). Busenberg and others (2001, p. 87) indicated that water in this well is mostly deep water, and there were no CFCs present so the well is not influenced by wastewater discharge. The variable concentrations may be due to response to above-average and below-average periods of recharge influencing the deeper part of the aquifer at this well location.

USGS 86.—Analyses of trends for constituents in water from this well indicate decreasing trends for chloride and nitrate concentrations, and no trend for sodium concentrations (table 3). Busenberg and others (2001, p. 92) indicated that the young fraction of water in this well was about 12 years old and was related to Big Lost River streamflow, so the decreasing chloride and nitrate concentrations could be due to dilution by Big Lost River water.

USGS 97.—Analyses of constituents in water from this well indicate increasing trends for chloride, sodium, nitrate, bromide, and molybdenum concentrations, a decreasing trend in sulfate and and barium concentrations, and no trend for arsenic (table 3). Busenberg and others (2001, p. 87) indicated that the young fraction of water in this well was about 6 years old and probably was related to wastewater disposal at the NRF, which could account for the variable and increasing concentration trends for the constituents in the well.

USGS 98.—Analyses of trends for constituents in water from this well indicate a decreasing trend for chloride concentrations, and no trends for most of the other constituents (table 3). Almost all constituents show a small decrease in concentration when the pump was lowered in 2005 (appendix C), which may account for the decreasing concentration trends. Busenberg and others (2001, p. 87) indicated that the young fraction of water in this well was about 7 years old and probably was related to wastewater disposal at the NRF, so decreasing chloride concentrations in wastewater could account for the decreasing chloride concentration trend.

USGS 99.—Analyses of constituents in water from this well indicate increasing trends for chloride and sodium concentrations and no trends for sulfate, bromide, and nitrate concentrations (table 3). Busenberg and others (2001, p. 88) indicated that the young fraction of water in this well was about 4 years old and probably was related to wastewater disposal at the NRF, which could account for the variable and increasing concentration trends for chloride and sodium

USGS 100.—Analyses of trends for constituents in water from this well indicate increasing trends for sodium and sulfate concentrations, and no trend for chloride concentrations (table 3). Busenberg and others (2001, p. 81) indicates that the water in this well is about 40–50 percent young recharge water, and the remainder is regional recharge water. The increasing trends may be due to regional anthropogenic influences on the aquifer.

USGS 101.—Analyses of trends for constituents in water from this well indicate increasing trends for chloride and nitrate concentrations, and no trend for sodium concentrations (table 3). Busenberg and others (2001, p. 81) indicated that water in this well is about 20–40 percent young recharge water, and the remainder is regional recharge water. The increasing trends may be due to regional anthropogenic influences on the aquifer.

USGS 102.—Analyses of constituents in water from this well indicate no trend but variable changes (appendix C) for chloride and sulfate and increasing trends for sodium, bromide, nitrate, orthophosphate, and TOC concentrations (tables 3 and 4). Busenberg and others (2001, p. 88) indicated that the young fraction of water in this well was about 6 years old and probably was related to wastewater disposal at the NRF, which could account for the variable changes and increasing concentration trends for the constituents.

USGS 103.—Analyses of trends for constituents in water from this well indicate no trends for chloride and sodium concentrations, and an increasing trend for nitrate concentrations (table 3). Busenberg and others (2001, p. 88) indicated that the young fraction of water in this well was about 26 years old and probably was related to wastewater disposal at INTEC, which could account for the increasing concentration trend for nitrate.

USGS 107.—Analyses of trends for constituents in water from this well indicate increasing trends for chloride, nitrate, and sodium concentrations (table 3). Busenberg and others (2001, p. 88) indicated that the young fraction of water in this well probably was older than 15 years. The increasing concentrations may be due to the overall increases occurring in the regional aquifer system due to agricultural practices northeast of the INL, or could be the result of wastewater disposal from facilities (possibly the PBF) finally reaching the well and elevating the concentrations.

USGS 109.—Analyses of trends for constituents in water from this well indicate a decreasing trend for chloride concentrations, no trend for sodium concentrations, and an increasing trend for nitrate concentrations (table 3). Busenberg and others (2001, p. 88–89) indicated that the young fraction of water in this well was about 18–20 years old and probably was related to wastewater disposal at INTEC, so variable concentrations in wastewater discharge may explain why chloride decreases and nitrate increases. The higher concentrations of chloride in the mid-1980s (appendix C) support this wastewater disposal theory.

USGS 110 and 110A.—Analyses of trends for constituents in water from this well (USGS 110) and the replacement well (USGS 110A, redrilled next to USGS 110) indicate an increasing trend for chloride concentrations in well USGS 110, but no trend for well USGS 110A; no trends for sodium concentrations in either well; and no trend for nitrate in well 110A (table 3). Busenberg and others (2001, p. 81) indicated that water in well USGS 110A is about half young recharge water, and the remainder is regional recharge water.

USGS 117.—Analyses of trends for constituents in water from this well indicate decreasing trends for chloride and nitrate concentrations, an increasing trend for sodium concentrations, and no trend for sulfate concentrations (table 3). Busenberg and others (2001, p. 89) indicated that although this well is downgradient of the RWMC, it does not contain CFCs and the young fraction of water probably is greater than 33 years old. It is not clear why the concentration trends for the different constituents are variable.

USGS 119.—Analyses of trends for constituents in water from this well indicate an increasing trend for sulfate concentrations, and no trends for chloride, sodium, and nitrate concentrations (table 3). Busenberg and others (2001, p. 89) could not date the young fraction of water because of contamination with CFCs. The increasing sulfate concentrations could be due to waste disposal practices either at the RWMC, or at the ATRC, but it is unclear why the other constituents do not show any trend.

USGS 121.—Analyses of trends for constituents in water from this well indicate a decreasing trend for chloride concentrations, and no trends for nitrate, sulfate, and sodium concentrations (table 3). Although overall chloride concentrations show a decreasing trend, concentrations decrease and increase with time. This well is located next to the Big Lost River; so above-average seepage and no seepage from the river may cause the variable concentrations. Busenberg and others (2001, p. 89) calculated an age of 15.5 years for the young fraction of water, and speculated that the young fraction was related to wastewater discharge at the NRF.

USGS 125.—Analyses of trends for constituents in water from this well indicate decreasing trends for chloride and sodium concentrations, and no trend for nitrate concentrations (table 3). Busenberg and others (2001, p. 90) indicated a recharge age of 17 years for the young fraction of water. They also attributed the young fraction to the INTEC disposal well, so decreasing trends for chloride and sodium concentrations (appendix C) could be representative of decreased discharge of wastewater constituents at the INTEC.

USGS 126A.—Analyses of trends for constituents in water from this well indicate no trends for chloride, nitrate, and sodium concentrations (table 3). Swanson and others (2003) indicated that water from this well could be modeled from upgradient wells in the Birch Creek basin. Because this water can be attributed to Birch Creek underflow, a reduced sampling frequency may be considered.

USGS 126B.—Analyses of trends for constituents in water from this well indicate no trends for chloride, nitrate, and sodium concentrations (table 3). Swanson and others (2003) indicated that water from this well could be modeled from upgradient wells in the Birch Creek basin. Because this water can be attributed to Birch Creek underflow, a reduced sampling frequency may be considered.

WS INEL 1.—Analyses of trends for constituents in water from this well indicate decreasing trends for bromide, chloride, nitrate, sodium, and sulfate concentrations, and no trend for TOC concentrations (tables 3 and 4). Busenberg and others (2001, p. 85) indicated difficulty with determining the young fraction of water in this well. A radioactive source was lost in the well after completion and initial sampling in 1978, and the source was cemented in the bottom of the well. Decreasing trends may not be representative of the aquifer conditions, but instead may be a remnant of contamination of the well from cementing in the source or from drilling fluids used to drill the nearby well INEL 1. Chloride, sodium, and sulfate all show increasing concentrations from the late 1990s until the early 2000s (appendix C). The variable concentrations could be due to a combination of wastewater disposal at the NRF and/or above-average and below-average periods of recharge to the aquifer, or from the previously listed contamination sources. The higher than background concentrations of bromide, chloride, nitrate, sodium, and sulfate seem related to contamination, and not associated with natural aquifer conditions.

Big Lost River at Experimental Dairy Farm near Howe, Idaho.—Analyses of trends for constituents in water from this site indicate no trend for chloride concentrations (table 3). However, concentrations of chloride tended to be higher during wetter periods, when there is a longer period of flow at the site, because the water has more time to acquire dissolved constituents.

Big Lost River below INL Diversion near Arco, Idaho.—Analyses of trends for constituents in water from this site indicate a decreasing trend for chloride concentrations (table 3); however, concentrations of chloride tended to be higher during wetter periods, when there is a longer period of flow at the site, because the water has more time to acquire dissolved constituents. Therefore, the decreasing trend may be associated more with recent, shorter periods of flow (2005, 2006, and 2009) consisting of mostly snowfall runoff with low chloride concentrations than with natural environmental influence on the river.

Big Lost River below Mackay Reservoir near Mackay, Idaho.—Analyses of constituents in water from this site indicate a decreasing trend for chloride concentrations (table 3). Because this site has continuous flow downstream of Mackay Dam, decreasing trends should be representative of system response to natural environmental influences and not associated with shorter periods of flow or periodic runoff. The chemical changes at this site generally appear be carried down throughout the other downgradient sites sampled.

Big Lost River near Arco, Idaho.—Analyses of trends for constituents in water from this site indicate no concentration trends for all constituents (table 3). Data for chloride from the 1990 to 2009 timeframe when the river was dry for about one-half the time indicates similar concentrations as other Big Lost River sites (Big Lost River below INL Diversion and Big Lost River near the Experimental Dairy Farm) (appendix C) that have periodic flow.

Birch Creek at Blue Dome Inn, near Reno, Idaho.—Analyses of trends for constituents in water from this site indicate a decreasing trend for chloride concentrations (table 3). Because this site has continuous flow from springs and runoff in the upper basin, decreasing trends should be representative of system response to natural environmental influences.

Little Lost River near Howe, Idaho.—Analyses of trends for constituents in water from this site indicate an increasing trend for chloride concentrations, and no trends for all other constituents (table 3). The increasing chloride concentrations could be due to agricultural influences in the Little Lost River basin.

Mud Lake near Terreton, Idaho.—Analyses of trends for constituents in water from this site indicate an increasing trend for chloride concentrations, and no trends for all other constituents (table 3). The increasing chloride concentrations could be due to agricultural influences in the Mud Lake basin.

Summary and Conclusions

The USGS, in cooperation with the U.S. Department of Energy, has maintained a water-quality monitoring network at the INL since 1949 to define the quality of water for human and industrial use and to better understand the location and movement of contaminants in the ESRP aquifer at the INL. Water-quality trends for 67 aquifer wells and 7 surface-water sites were examined for selected constituents, including major cations, anions, nutrients, trace elements, and total organic carbon for data collected from 1949 through 2009. Water-quality trends were determined using (1) the nonparametric Kendall's *tau* correlation coefficient, *p*-value, Theil-Sen slope estimator, and summary statistics for uncensored data; and (2) the Kaplan-Meier method for calculating summary statistics, Kendall's *tau* correlation coefficient, *p*-value, and Akritas-Theil-Sen slope estimator for robust linear regression for censored data. Water-quality trends were determined to assist with future management decisions concerning which wells to sample and which constituent types to monitor.

Water-quality trends were determined from groundwater for the period of record after the date of pump installations, or after 1980 through 2009, to eliminate variability associated with sample-collection methods and data not collected under a quality assurance program. Outlier data that did not correlate with field-specific conductance measurements and replicate values were not used. Water-quality trends were not determined for the radiochemical constituents analyzed because almost all concentrations were less than the reporting level.

Trends for field measurements of pH, specific conductance, and water temperature were analyzed. It was difficult to assess whether or not the trends for pH were due to changing aquifer conditions or because of the variability throughout the history of the sample collection methods or meter types. Trends of specific conductance in water from a few wells were affected by collection method. Trends of specific conductance in water for most of the other sites generally were similar to trends for the chemical constituents chloride, sodium, and sulfate. For most of the sites, water temperature did not show a statistical trend.

Chloride results indicate that groundwater influenced by seepage loss from the Big Lost River have decreasing trends or, in some cases, have trends that correlate with periods of above-average recharge and no recharge from the river. Water samples from four sampling locations on the Big Lost River show a decreasing trend or no trend for chloride concentrations, and concentrations generally are much less than those in the aquifer. Above-average and below-average periods of recharge also affect concentration trends for sodium, sulfate, nitrate, and a few trace elements in several wells. Analyses of trends for constituents in water from several of the wells that is mostly regionally derived groundwater generally show an increasing trend in chloride, sodium, sulfate, and nitrate concentrations. These increases are attributed to agricultural or other anthropogenic influences upgradient of the INL.

Chemical constituent trends in water from several wells near the NRF may be influenced by wastewater disposal at the facility or could be due to anthropogenic influence from the Little Lost River basin. Increasing sodium concentrations in a few wells near the ATRC and INTEC also could be affected by wastewater discharge, but trends of other constituents do not show a similar correlation. Three wells located at and downgradient from the PBF area (SPERT 1, Site 9, and USGS 107) all show increasing trends of several constituents. The increases could be due to wastewater disposal in the PBF area; more wells may need to be added to the sample program to fully characterize that area.

Water from several wells located in the southwestern part of the INL and southwest of the INL may show concentration trends for chloride and sodium that were influenced by wastewater disposal at INTEC and ATRC, even though these wells were selected for this study because it was believed they were not influenced. Some constituents show decreases that could be attributed to the decreasing constituent concentrations in the wastewater from the late 1970s to 2009, and from changes in wastewater disposal practices. Ages of the young recharge water in many of the wells are more than 20 years old, so samples collected in the early 1990s are more representative of water discharged in the 1960s and 1970s, when concentrations of chloride, nitrate, sodium, and sulfate in wastewater were much higher. Water sampled in 2009 would be representative of the low concentrations of chloride and sodium in wastewater discharged in the late 1980s. Several wells in the central and southern part of the aquifer at the INL show increased sodium concentrations, and in most cases, the concentrations were less than what was considered as the background concentration for the aquifer in 1974. Many wells are open to large mixed sections of the aquifer, and increasing concentration trends may indicate that the long history of wastewater disposal in the central part of the INL is causing sodium concentrations to increase.

At several sites, a reduced sampling frequency may be considered because all the constituents showed no trend. Water samples collected for radiochemical constituents, fluoride, some trace elements, total organic carbon, and volatile organic compounds could be discontinued if trends continue as they currently (2009) are presented.

References Cited

Ackerman, D.J., 1991a, Transmissivity of the Snake River Plain aquifer at the Idaho National Engineering Laboratory, Idaho: U.S. Geological Survey Water-Resources Investigations Report 91–4058 (DOE/ID–22097), 35 p. (Also available at http://pubs.er.usgs.gov/publication/wri914058.)

Ackerman, D.J., 1991b, Transmissivity of perched aquifers at the Idaho National Engineering Laboratory, Idaho: U.S. Geological Survey Water-Resources Investigations Report 91-4114 (DOE/ ID-22099), 27 p. (Also available at http://pubs.er.usgs.gov/usgspubs/wri/wri914114.)

Ackerman, D.J., Rattray, G.W., Rousseau, J.P., Davis, L.C., and Orr, B.R., 2006, A conceptual model of ground-water flow in the eastern Snake River Plan aquifer at the Idaho National Laboratory and vicinity with implications for contaminant transport: U.S. Geological Survey Scientific Investigations Report 2006–5122, 62 p. (Also available at http://pubs.er.usgs.gov/publication/sir20065122.)

Ackerman, D.J., Rousseau, J.P., Rattray, G.W., and Fisher, J.C., 2010, Steady-state and transient models of groundwater flow and advective transport, eastern Snake River Plain aquifer, Idaho National Laboratory and vicinity, Idaho: U.S. Geological Survey Scientific Investigations Report 2010–5123 (DOE/ID-22209), 220 p. (Also available at http://pubs.er.usgs.gov/publication/sir20105123.)

Anderson, S.R., 1991, Stratigraphy of the unsaturated zone and uppermost part of the Snake River Plain aquifer at the Idaho Chemical Processing Plant and Test Reactors Area, Idaho National Engineering Laboratory, Idaho: U.S. Geological Survey Water-Resources Investigations Report 91-4010 (DOE/ID–22095), 71 p. (Also available at http://pubs.er.usgs.gov/publication/wri914010.)

Anderson, S.R., and Lewis, B.D., 1989, Stratigraphy of the unsaturated zone at the Radioactive Waste Management Complex, Idaho National Engineering Laboratory, Idaho: U.S. Geological Survey Water-Resources Investigations Report 89-4065 (DOE/ID-22080), 54 p. (Also available at http://pubs.er.usgs.gov/publication/wri894065.)

Barraclough, J.T., and Jensen, R.G., 1976, Hydrologic data for the Idaho National Engineering Laboratory site, Idaho, 1971 to 1973: U.S. Geological Survey Open-File Report 75–318 (IDO–22055), 52 p. (Also available at http://pubs.er.usgs.gov/publication/ofr75318.)

Barraclough, J.T., Lewis, B.D., and Jensen, R.G., 1981, Hydrologic conditions at the Idaho National Engineering Laboratory, Idaho, emphasis 1974–1978: U.S. Geological Survey Water-Resources Investigations Open-File Report 81–526 (IDO-22060), 122 p. (Also available at http://pubs.er.usgs.gov/publication/ofr81526.)

Barraclough, J.T., Robertson, J.B., and Janzer, V.J., 1976, Hydrology of the solid waste burial ground as related to the potential migration of radionuclides, Idaho National Engineering Laboratory, with a section on drilling and sample analyses, by L.G. Saindon: U.S. Geological Survey Open-File Report 76–471 (IDO–22056), 183 p. (Also available at http://pubs.er.usgs.gov/publication/ofr76471.)

Barraclough, J.T., Teasdale, W.E., and Jensen, R.G., 1967a, Hydrology of the National Reactor Testing Station, Idaho, 1965: U.S. Geological Survey Open-File Report (IDO–22048), 107 p.

Barraclough, J.T., Teasdale, W.E., Robertson, J.B., and Jensen, R.G., 1967b, Hydrology of the National Reactor Testing Station, Idaho, 1966: U.S. Geological Survey Open-File Report 67–12 (IDO–22049), 98 p. (Also available at http://pubs.er.usgs.gov/publication/ofr6712.)

Bartholomay, R.C., 1993, Concentrations of tritium and strontium-90 in water from selected wells at the Idaho National Engineering Laboratory after purging one, two, and three borehole volumes: U.S. Geological Survey Water-Resources Investigations Report 93–4201 (DOE/ID-22111), 21 p. (Also available at http://pubs.er.usgs.gov/usgspubs/wri/wri934201.)

Bartholomay, R.C., 1998, Distribution of selected radiochemical and chemical constituents in perched ground water, Idaho National Engineering Laboratory, Idaho, 1992–95: U.S. Geological Survey Water-Resources Investigations Report 98–4026 (DOE/ID–22145), 59 p. (Also available at http://pubs.er.usgs.gov/publication/wri984026.)

Bartholomay, R.C., 2009, Iodine-129 in the Snake River Plain aquifer at and near the Idaho National Laboratory, Idaho, 2003 and 2007: U.S. Geological Survey Scientific Investigations Report 2009–5088, 28 p. (Also available at http://pubs.er.usgs.gov/publication/sir20095088.)

Bartholomay, R.C., Knobel, L.L., and Rousseau, J.P., 2003, Field methods and quality-assurance plan for quality-of-water activities, U.S. Geological Survey, Idaho National Engineering and Engineering Laboratory, Idaho: U.S. Geological Survey Open-File Report 2003-42 (DOE/ID–22182), 45 p. (Also available at http://pubs.er.usgs.gov/publication/ofr0342.)

Bartholomay, R.C., Orr, B.R., Liszewski, M.J., and Jensen, R.G., 1995, Hydrologic conditions and distribution of selected radiochemical and chemical constituents in water, Snake River Plain aquifer, Idaho National Engineering Laboratory, Idaho, 1989 through 1991: U.S. Geological Survey Water-Resources Investigations Report 95–4175 (DOE/ID–22123), 47 p. (Also available at http://pubs.er.usgs.gov/publication/wri954175.)

Bartholomay, R.C., and Tucker, B.J., 2000, Distribution of selected radiochemical and chemical constituents in perched ground water, Idaho National Engineering and Environmental Laboratory, Idaho, 1996–98: U.S. Geological Survey Water-Resources Investigations Report 2000–4222 (DOE/ID–22168), 51 p. (Also available at http://pubs.er.usgs.gov/publication/wri004222.)

Bartholomay, R.C., Tucker, B.J., Ackerman, D.J., and Liszewski, M.J., 1997, Hydrologic conditions and distribution of selected radiochemical and chemical constituents in water, Snake River Plain aquifer, Idaho National Engineering Laboratory, Idaho, 1992 through 1995: U.S. Geological Survey Water-Resources Investigations Report 97–4086 (DOE/ID–22137), 57 p. (Also available at http://pubs.er.usgs.gov/publication/wri974086.)

Bartholomay, R.C., Tucker, B.J., Davis, L.C., and Green, M.R., 2000, Hydrologic conditions and distribution of selected constituents in water, Snake River Plain aquifer, Idaho National Engineering and Environmental Laboratory, Idaho, 1996 through 1998: U.S. Geological Survey Water-Resources Investigations Report 2000–4192 (DOE/ID–22167), 52 p. (Also available at http://pubs.er.usgs.gov/publication/wri004192.)

Bartholomay, R.C., and Williams, L.M., 1996, Evaluation of preservation methods for selected nutrients in ground water at the Idaho National Engineering Laboratory, Idaho: U.S. Geological Survey Water-Resources Investigations Report 96–4260 (DOE/ID–22131), 16 p. (Also available at http://pubs.er.usgs.gov/publication/wri964260.)

Bodnar, L.Z., and Percival, D.R., eds., 1982, Analytical Chemistry Branch procedures manual-Radiological and Environmental Sciences Laboratory: U.S. Department of Energy Report IDO-12096 [variously paged].

Busenberg, Eurybiades, Plummer, L.N., and Bartholomay, R.C., 2001, Estimated age and source of the young fraction of ground water at the Idaho National Engineering and Environmental Laboratory: U.S. Geological Survey Water-Resources Investigations Report 2001–4265 (DOE/ID-22177), 144 p. (Also available at http://pubs.er.usgs.gov/publication/wri014265.)

Cecil, L.D., Orr, B.R., Norton, T.J., and Anderson, S.R., 1991, Formation of perched ground-water zones and concentrations of selected chemical constituents in water, Idaho National Engineering Laboratory, Idaho, 1986–88: U.S. Geological Survey Water-Resources Investigations Report 91–4166 (DOE/ID–22100), 53 p. (Also available at http://pubs.er.usgs.gov/publication/wri914166.)

Cecil, L.D., Welhan, J.A., Green, J.R., Frape, S.K., and Sudicky, E.R., 2000, Use of chlorine-36 to determine regional-scale aquifer dispersivity, eastern Snake River Plain aquifer, Idaho/USA: Nuclear Instruments and Methods in Physics Research, section B, v. 172, issues 1–4, p. 679–687.

Childress, C.J.O., Forman, W.T., Connor, B.F., and Maloney, T.J., 1999, New reporting procedures based on long-term method detection levels and some considerations for interpretations of water-quality data provided by the U.S. Geological Survey National Water Quality Laboratory: U.S. Geological Survey Open-File Report 99–193, 19 p. (Also available at http://pubs.er.usgs.gov/publication/ofr99193.)

Claassen, H.C., 1982, Guidelines and techniques for obtaining water samples that accurately represent the water chemistry of an aquifer: U.S. Geological Survey Open-File Report 82–1024, 49 p. (Also available at http://pubs.er.usgs.gov/publication/ofr821024.)

Currie, L.A., 1984, Lower limit of detection-definition and elaboration of a proposed position for radiological effluent and environmental measurements: U.S. Nuclear Regulatory Commission NUREG/CR–4007, 139 p.

Davis, L.C., 2006a, An update of the distribution of selected radiochemical and chemical constituents in perched ground water, Idaho National Laboratory, Idaho, emphasis 1999–2001: U.S. Geological Survey Scientific Investigations Report 2006–5236 (DOE/ID-22199), 48 p. (Also available at http://pubs.er.usgs.gov/publication/sir20065236.)

Davis, L.C., 2006b, An update of hydrologic conditions and distribution of selected constituents in water, Snake River Plain aquifer, Idaho National Laboratory, Idaho, emphasis 1999–2001: U.S. Geological Survey Scientific Investigations Report 2006–5088 (DOE/ID-22197), 48 p. (Also available at http://pubs.er.usgs.gov/publication/sir20065088.)

Davis, L.C., 2008, An update of hydrologic conditions and distribution of selected constituents in water, Snake River Plain aquifer and perched-water zones, Idaho National Laboratory, Idaho, emphasis 2002–05: U.S. Geological Survey Scientific Investigations Report 2008–5089 (DOE/ID-22203), 75 p. (Also available at http://pubs.er.usgs.gov/publication/sir20085089.)

Davis, L.C., 2010, An update of hydrologic conditions and distribution of selected constituents in water, Snake River Plain aquifer and perched groundwater zones, Idaho National Laboratory, Idaho, emphasis 2006–08: U.S. Geological Survey Scientific Investigations Report 2010–5197 (DOE/ID-22212), 80 p. (Also available at http://pubs.er.usgs.gov/publication/sir20105197.)

Duke, C.L., Roback, R.C., Reimus, P.W., Bowman, R.S., McLing, T.L., Baker, K.E., and Hull, L.C., 2007, Elucidation of flow and transport processes in a variably saturated system of interlayered sediment and fractured rock using tracer tests: Vadose Zone Journal, v. 6, no. 4, p. 855–867.

Faires, L.M., 1993, Methods of analysis by the U.S. Geological Survey National Water Quality Laboratory—Determinations of metals in water by inductively coupled plasma-mass spectrometry: U.S. Geological Survey Open-File Report 92–634, 28 p. (Also available at http://pubs.er.usgs.gov/publication/ofr92634.)

Fisher, J.C., and Davis, L.C., 2011, Analysis of data collected from a monitoring network, Package 'Trends': version 0.1-2, accessed August 6, 2012, at https://github.com/jfisher-usgs/Trends.

Fishman, M.J., ed., 1993, Methods of analysis by the U.S. Geological Survey National Water Quality Laboratory—Determination of inorganic and organic constituents in water and fluvial sediments: U.S. Geological Survey Open-File Report 93–125, 217 p. (Also available at http://pubs.er.usgs.gov/publication/ofr93125.)

Fishman, M.J., and Friedman, L.C., eds., 1989, Methods for determination of inorganic substances in water and fluvial sediments (3d ed.): U.S. Geological Survey Techniques of Water-Resources Investigations, book 5, chap. A1, 545 p. (Also available at http://pubs.er.usgs.gov/publication/twri05A1.)

French, D.L., Tallman, R.E., and Taylor, K.A., 1999, Idaho National Engineering Laboratory nonradiological waste management information for 1998 and record-to-date: U.S. Department of Energy, Waste Management Programs Division, Idaho Operations Office Publication, DOE/ID-10057 (98) [variously paged].

Garabedian, S.P., 1986, Application of a parameter-estimation technique to modeling the regional aquifer underlying the eastern Snake River Plain, Idaho: U.S. Geological Survey Water Supply Paper 2278, 60 p. (Also available at http://pubs.er.usgs.gov/publication/wsp2278.)

Goerlitz, D.F., and Brown, Eugene, 1972, Methods for analysis of organic substances in water: U.S. Geological Survey Techniques of Water-Resources Investigations, book 5, chap. A3, 40 p. (Also available at http://pubs.er.usgs.gov/publication/twri05A3_1972.)

Helsel, D.R., 2005, Nondetects and data analysis—Statistics for censored environmental data: Hoboken, N.J., Wiley, 250 p.

Hem, J.D., 1989, Study and interpretation of the chemical characteristics of natural water (3d ed.): U.S. Geological Survey Water-Supply Paper 2254, 263 p. (Also available at http://pubs.er.usgs.gov/publication/wsp2254.)

Hull, L.C., 1989, Conceptual model and description of the affected environment for the TRA warm waste pond (Waste Management Unit TRA-03): EG and G Idaho Informal Report EGG–ER–8644, 125 p.

Jones, P.H., 1961, Hydrology of waste disposal, National Reactor Testing Station, Idaho, an interim report: U.S. Atomic Energy Commission, Idaho Operations Office Publication IDO–22042–USGS, 152 p.

Knobel, L.L., 2006, Evaluation of well-purging effects on water-quality results for samples collected from the eastern Snake River Plain aquifer underlying the Idaho National Laboratory, Idaho: U.S. Geological Survey Scientific Investigations Report 2006–5232 (DOE/ID-22200), 52 p. (Also available at http://pubs.er.usgs.gov/publication/sir20065232.)

Knobel, L.L., Tucker, B.J. and Rousseau, J.P., 2008, Field methods and quality-assurance plan for quality-of-water activities, U.S. Geological Survey, Idaho National Laboratory, Idaho: U.S. Geological Survey Open-File Report 2008–1165 (DOE/ID–22206), 36 p. (Also available at http://pubs.er.usgs.gov/publication/ofr20081165.)

Lee, Lopaka, 2012, Nada—Nondetects and data analysis for environmental data; The R Project for Statistical Computing Web site accessed May 11, 2012, at http://cran.r-project.org/web/packages/NADA/index.html.

Lewis, B.D., and Jensen, R.G., 1985, Hydrologic conditions at the Idaho National Engineering Laboratory, Idaho, 1979–1981 update: U.S. Geological Survey Hydrologic Investigations Atlas HA–674, 2 sheets. (Also available at http://pubs.er.usgs.gov/publication/ha674.)

Mann, L.J., 1986, Hydraulic properties of rock units and chemical quality of water for INEL-1—A 10,365-foot deep test hole drilled at the Idaho National Engineering Laboratory, Idaho: U.S. Geological Survey Water-Resources Investigations Report 86–4020 (DOE/ID–22070), 23 p. (Also available at http://pubs.er.usgs.gov/publication/wri864020.)

Mann, L.J., 1996, Quality-assurance plan and field methods for quality-of-water activities, U.S. Geological Survey, Idaho National Engineering Laboratory, Idaho: U.S. Geological Survey Open-File Report 96–615 (DOE/ID-22132), 37 p. (Also available at http://pubs.er.usgs.gov/publication/ofr96615.)

Mann, L.J., and Beasley, T.M., 1994, Iodine-129 in the Snake River Plain aquifer at and near the Idaho National Engineering Laboratory, Idaho, 1990–1991: U.S. Geological Survey Water-Resources Investigations Report 94-4053 (DOE/ID-22115), 27 p. (Also available at http://pubs.er.usgs.gov/publication/wri944053.)

McCurdy, D.E., Garbarino, J.R., and Mullin, A.H., 2008, Interpreting and reporting radiological water-quality data: U.S. Geological Survey Techniques and Methods, book 5, chap. B6, 33 p. (Also available at http://pubs.er.usgs.gov/publication/tm5B6.)

Morris, D.A., Barraclough, J.T., Chase, G.H., Teasdale, W.E., and Jensen, R.G., 1965, Hydrology of subsurface waste disposal, National Reactor Testing Station, Idaho, annual progress report, 1964: U.S. Atomic Energy Commission, Idaho Operations Office Publication, IDO–22047–USGS, 186 p.

Morris, D.A., Barraclough, J.T., Hogenson, G.M., Shuter, Eugene, Teasdale, W.E., Ralston, D.A., and Jensen, R.G., 1964, Hydrology of subsurface waste disposal, National Reactor Testing Station, Idaho, annual progress report, 1963: U.S. Atomic Energy Commission, Idaho Operations Office Publication, IDO–22046–USGS, 97 p.

Morris, D.A., Hogenson, G.M., Shuter, Eugene, and Teasdale, W.E., 1963, Hydrology of waste disposal, National Reactor Testing Station, Idaho, annual progress report, 1962: U.S. Atomic Energy Commission, Idaho Operations Office Publication, IDO–22044–USGS, 99 p.

Nace, R.L., Voegeli, P.T., Jones, J.R., and Deutsch, Morris, 1975, Generalized geologic framework of the National Reactor Testing Station, Idaho: U.S. Geological Survey Professional Paper 725-B, 48 p. (Also available at http://pubs.er.usgs.gov/publication/pp725B.)

Nimmo, J.R., Perkins, K.S., Rose, P.A., Rousseau, J.P., Orr, B.R., Twining, B.V., and Anderson, S.R., 2002, Kilometer-scale rapid transport of naphthalene sulfonate tracer in the unsaturated zone at the Idaho National Engineering and Environmental Laboratory: Vadose Zone Journal, v. 1, issue 1, p. 89–101.

Olmsted, F.H., 1962, Chemical and physical character of ground water in the National Reactor Testing Station, Idaho: U.S. Atomic Energy Commission, Idaho Operations Office Publication IDO-22043-USGS, 142 p.

Orr, B.R., 1999, A transient numerical simulation of perched ground-water flow at the Test Reactor Area, Idaho National Engineering and Environmental Laboratory, Idaho, 1952–94: U.S. Geological Survey Water-Resources Investigations Report 99–4277 (DOE/ID–22162), 54 p. (Also available at http://pubs.er.usgs.gov/publication/wri994277.)

Orr, B.R., and Cecil, L.D., 1991, Hydrologic conditions and distribution of selected chemical constituents in water, Snake River Plain aquifer, Idaho National Engineering Laboratory, Idaho, 1986 to 1988: U.S. Geological Survey Water-Resources Investigations Report 91–4047 (DOE/ID–22096), 56 p. (Also available at http://pubs.er.usgs.gov/publication/wri914047.)

Orr, B.R., Cecil, L.D., and Knobel, L.L., 1991, Background concentrations of selected radionuclides, organic compounds, and chemical constituents in ground water in the vicinity of the Idaho National Engineering Laboratory: U.S. Geological Survey Water-Resources Investigations Report 91–4015 (DOE/ID–22094), 52 p. (Also available at http://pubs.er.usgs.gov/publication/wri914015.)

Pittman, J.R., Fischer, P.R., and Jensen, R.G., 1988, Hydrologic conditions at the Idaho National Engineering Laboratory, 1982 to 1985: U.S. Geological Survey Water-Resources Investigations Report 89–4008 (DOE/ID–22078), 73 p. (Also available at http://pubs.er.usgs.gov/publication/wri894008.)

Plummer, L.N., Rupert, M.G., Busenberg, Eurybiades, and Schlosser, P., 2000, Age of irrigation water in ground water from the eastern Snake River Plain aquifer, South-central Idaho: Ground Water, v. 38, no. 2, p. 264–283.

Pritt, J. W., 1989, Quality assurance of sample containers and preservatives at the U.S. Geological Survey National Water Quality Laboratory, in Pederson, G.L., and Smith, M.M., compilers, U.S. Geological Survey Second National Symposium on Water Quality—Abstracts of the Technical Sessions: U.S. Geological Survey Open-File Report 89–409, 111 p. (Also available at http://pubs.er.usgs.gov/publication/ofr89409.)

R Development Core Team, 2011, A language and environment for statistical computing, ISBN 3-900051-07-0: The R Project for Statistical Computing Web site, accessed May 11, 2012, at http://www.R-project.org.

Robertson, J.B., 1976, Numerical modeling of subsurface radioactive solute transport from waste seepage ponds at the Idaho National Engineering Laboratory: U.S. Geological Survey Open-File Report 76–717 (IDO-22057), 68 p. (Also available at http://pubs.er.usgs.gov/usgspubs/ofr/ofr76717.)

Robertson, J.B., Schoen, Robert, and Barraclough, J.T., 1974, The influence of liquid waste disposal on the geochemistry of water at the National Reactor Testing Station, Idaho, 1952–1970: U.S. Geological Survey Open-File Report 73–238 (IDO–22053), 231 p. (Also available at http://pubs.er.usgs.gov/publication/ofr73238.)

Schramke, J. A., Murphy, E.M., and Wood, B. D., 1996, The use of geochemical mass-balance and mixing models to determine groundwater sources, Applied Geochemistry, v. 11, issue 4, p. 523–539.

Sehlke, Gerald, and Bickford, F.E., 1993, Idaho National Engineering Laboratory ground-water monitoring plan: E G and G Idaho, Inc., and Golder Associates, Inc., DOE/ID–10441, v. 1–2 [variously paged].

Stevens, H.H., Ficke, J.F., and Smoot, G.F., 1975, Water temperature—Influential factors, field measurement, and data presentation: U.S. Geological Survey Techniques of Water-Resources Investigations, book 1, chap. D1, 65 p. (Also available at http://pubs.er.usgs.gov/publication/twri01D1.)

Swanson, S.A., Rosentreter, J.J., Bartholomay, R.C., and Knobel, L.L., 2002, Geochemistry of the Little Lost River drainage basin, Idaho: U.S. Geological Survey Water-Resources Investigations Report 2002–4120 (DOE/ID-22179), 29 p. (Also available at http://pubs.er.usgs.gov/publication/wri024120.)

Swanson, S.A., Rosentreter, J.J., Bartholomay, R.C., and Knobel, L.L., 2003, Geochemistry of the Birch Creek drainage basin, Idaho: U.S. Geological Survey Water-Resources Investigations Report 2003–4272 (DOE/ID-22188), 36 p.(Also available at http://pubs.er.usgs.gov/publication/wri034272.)

Thatcher, L.L., Janzer, V.J., and Edwards, K.W., 1977, Methods for determination of radioactive substances in water and fluvial sediments: U.S. Geological Survey Techniques of Water-Resources Investigations, book 5, chap. A5, 95 p. (Also available at http://pubs.er.usgs.gov/publication/twri05A5.)

Timme, P.J., 1995, National Water Quality Laboratory, 1995 services catalog: U.S. Geological Survey Open-File Report 95–352, 120 p. (Also available at http://pubs.er.usgs.gov/publication/ofr95352.)

Tucker, B.J., and Orr, B.R., 1998, Distribution of selected radiochemical and chemical constituents in perched ground water, Idaho National Engineering Laboratory, Idaho, 1989–91: U.S. Geological Survey Water-Resources Investigations Report 98–4028 (DOE/ID–22144), 62 p. (Also available at http://pubs.er.usgs.gov/publication/wri984028.)

U.S. Department of Energy, 1995, Radiochemistry manual, revision 10: Idaho Falls, Idaho, U.S. Department of Energy, Radiological and Environmental Sciences Laboratory [variously paged].

U.S. Department of Energy, 2010, Waste Area Group 10, operable unit 10-08, annual monitoring status report for fiscal year 2009: U.S. Department of Energy, DOE/ID-11417, revision 0, unpaginated.

U.S. Environmental Protection Agency, 2001, Protection of environment—Code of Federal Regulations 40: Washington, D.C., Office of the Federal Register, National Archives and Records Administration, parts 136 to 149, 833 p.

U.S. Geological Survey, 1985, National water summary, 1984—Hydrologic events, selected water-quality trends, and ground-water resources: U.S. Geological Survey Water-Supply Paper 2275, 467 p. (Also available at http://pubs.er.usgs.gov/publication/wsp2275.)

U.S. Geological Survey, variously dated, National field manual for the collection of water-quality data: U.S. Geological Survey Techniques of Water-Resources Investigations, book 9, chaps. A1–A9. (Also available at http://pubs.er.usgs.gov/publication/twri09.)

Walker, F.W., Parrington, J.R., and Feiner, Frank, 1989, Nuclides and isotopes, chart of the nuclides (14th ed.): General Electric Company, Nuclear Energy Operations, 57 p.

Wegner, S.J., 1989, Selected quality assurance data for water samples collected by the U.S. Geological Survey, Idaho National Engineering Laboratory Idaho, 1980 to 1988: U.S. Geological Survey Water-Resources Investigations Report 89–4168 (DOE/ID–22085), 91 p. (Also available at http://pubs.er.usgs.gov/publication/wri894168.)

Wershaw, R.L., Fishman, M.J., Grabbe, R.R., and Lowe, L.E., eds., 1987, Methods for the determination of organic substances in water and fluvial sediments (revised ed.): U.S. Geological Survey Techniques of Water-Resource Investigation, book 5, chap. A3, 80 p. (Also available at http://pubs.er.usgs.gov/publication/twri05A3.)

Wilcox, R.R., 2012, Modern statistics for the social and behavioral sciences: A practical introduction: New York, CRC Press, 862 p.

Williams, L.M., 1996, Evaluation of quality assurance/quality control data collected by the U.S. Geological Survey for water-quality activities at the Idaho National Engineering Laboratory, Idaho, 1989 through 1993: U.S. Geological Survey Water-Resources Investigations Report 96–4148 (DOE/ID–22129), 116 p. (Also available at http://pubs.er.usgs.gov/publication/wri964148.)

Williams, L.M., 1997, Evaluation of quality assurance/quality control data collected by the U.S. Geological Survey for water-quality activities at the Idaho National Engineering Laboratory, Idaho, 1994 through 1995: U.S. Geological Survey Water-Resources Investigations Report 97–4058 (DOE/ID–22136), 87 p. (Also available at http://pubs.er.usgs.gov/publication/wri974058.)

Wood, W.W., 1976, Guidelines for collection and field analysis of ground-water samples for selected unstable constituents: U.S. Geological Survey Techniques of Water-Resource Investigation, book 1, chap. D2, 24 p. (Also available at http://pubs.er.usgs.gov/publication/twri01D2.)

Table 3. Statistical summaries and trend analyses of uncensored water-quality results for selected constituents in water from wells and surface-water sites at and near the Idaho National Laboratory, Idaho, 1965–2009.

[**Local name:** Local well identifier used in this study. **Site identifier:** Unique numerical identifiers used to access well data (http://waterdata.usgs.gov/nwis). **Constituent:** Arsenic, barium, chromium, manganese, molybdenum, uranium, and zinc in micrograms per liter. Other constituents in milligrams per liter. P, phosphorus; N, nitrogen. **Trend:** +, increasing; −, decreasing]

Local name	Site identifier	Constituent	Start date	End date	Sample size	Mean	Median	Minimum	Maximum	Standard deviation	p value	Slope	Trend
				Wells									
ANP 6	435152112443101	Chloride	01-01-1986	12-31-2009	22	16.8	16.6	11.0	21.8	3.24	0.037	0.325	+
		Sodium	01-01-1986	12-31-2009	20	10.2	10.2	9.20	11.3	0.497	0.045	0.035	+
		Sulfate	01-01-1986	12-31-2009	15	32.7	32.7	31.4	34.1	0.852	0.020	-0.114	−
		Nitrate + Nitrite as N	01-01-1986	12-31-2009	20	0.847	0.847	0.680	0.991	0.079	0.234	-0.006	no trend
		Orthophosphate as P	01-01-1986	12-31-2009	20	0.022	0.021	0.010	0.034	0.006	0.219	0.001	no trend
ANP 9	434856112400001	Chloride	01-01-1994	12-31-2009	23	12.2	12.1	11.1	13.0	0.512	0.775	0.004	no trend
		Sodium	01-01-1994	12-31-2009	23	14.5	14.5	13.6	15.5	0.464	0.987	0.000	no trend
		Nitrate + Nitrite as N	01-01-1994	12-31-2009	24	0.771	0.748	0.697	1.20	0.102	0.075	0.003	no trend
		Arsenic	01-01-1994	12-31-2009	23	2.30	2.28	1.62	3.00	0.374	0.538	0.011	no trend
		Barium	01-01-1994	12-31-2009	23	84.3	84.0	76.8	91.0	2.99	0.244	-0.144	no trend
		Chromium	01-01-1994	12-31-2009	23	3.90	2.92	1.93	17.1	3.05	0.000	-0.162	−
		Molybdenum	01-01-1994	12-31-2009	23	2.93	2.94	2.55	3.03	0.103	0.002	-0.010	−
		Uranium	01-01-1994	12-31-2009	23	2.23	2.24	2.00	2.53	0.178	0.000	0.029	+
		Zinc	01-01-1994	12-31-2009	23	10.7	9.58	1.16	64.0	12.5	0.010	-0.808	−
ARBOR Test	433509112384801	Chloride	01-01-1989	12-31-2009	29	14.6	14.6	3.0	18.0	0.956	0.311	0.023	no trend
AREA 2	433223112470201	Chloride	01-01-1992	12-31-2009	18	16.3	16.5	7.98	19.0	2.18	0.200	-0.045	no trend
		Sodium	01-01-1992	12-31-2009	18	15.6	16.0	9.36	17.0	1.99	0.427	-0.017	no trend
		Sulfate	01-01-1992	12-31-2009	15	17.8	17.3	15.4	28.2	3.02	0.007	0.177	+
		Nitrate + Nitrite as N	01-01-1992	12-31-2009	18	1.16	1.17	0.515	1.50	0.209	0.202	0.013	no trend
Atomic City	432638112484101	Chloride	01-01-1980	12-31-2009	50	17.0	17.0	13.0	22.0	1.61	0.715	0.000	no trend
		Sodium	01-01-1980	12-31-2009	28	15.6	16.0	7.00	18.1	2.53	0.000	0.091	+
Badging Facility Well	433042112535101	Chloride	01-01-1985	12-31-2009	33	16.6	16.0	14.0	23.0	1.85	0.571	-0.025	no trend
		Sodium	01-01-1985	12-31-2009	25	9.75	9.90	8.00	11.0	0.736	0.339	0.021	no trend
		Sulfate	01-01-1985	12-31-2009	16	20.8	20.8	20.0	22.0	0.699	0.252	0.044	no trend
		Nitrate + Nitrite as N	01-01-1985	12-31-2009	22	0.687	0.696	0.600	0.750	0.042	0.000	-0.005	−
CPP 4	433440112554401	Chloride	01-01-1980	12-31-2009	51	17.0	17.0	12.0	29.2	2.90	0.469	-0.059	no trend
		Sodium	01-01-1980	12-31-2009	47	7.52	7.66	6.00	8.40	0.496	0.010	0.026	+
		Nitrate + Nitrite as N	01-01-1980	12-31-2009	21	1.10	0.900	0.718	5.29	0.969	0.083	-0.010	no trend
EBR 1	433051113002601	Chloride	01-01-1980	12-31-2009	49	6.92	6.60	5.00	19.0	1.88	0.826	0.000	no trend
		Sodium	01-01-1980	12-31-2009	37	8.10	8.48	3.00	9.50	1.24	0.022	0.049	+
		Nitrate + Nitrite as N	01-01-1980	12-31-2009	30	0.417	0.381	0.270	0.900	0.136	0.496	0.001	no trend
Fire Station 2	433548112562301	Chloride	01-01-1980	12-31-1996	54	16.6	16.0	11.0	27.0	3.28	0.344	-0.145	no trend
		Sodium	01-01-1980	12-31-1996	19	8.02	8.00	6.00	10.0	0.853	0.087	0.063	no trend
		Nitrate + Nitrite as N	01-01-1980	12-31-1996	12	1.17	1.10	0.790	1.50	0.179	0.047	0.035	+
Highway 3	433256113002501	Chloride	01-01-1980	12-31-2009	50	6.31	6.00	4.00	12.0	1.16	0.093	-0.018	no trend
		Sodium	01-01-1980	12-31-2009	35	5.74	5.89	4.00	6.60	0.511	0.554	0.007	no trend
		Nitrate + Nitrite as N	01-01-1980	12-31-2009	29	0.373	0.372	0.296	0.570	0.060	0.003	0.004	+

Table 3   41

Table 3. Statistical summaries and trend analyses of uncensored water-quality results for selected constituents in water from wells and surface-water sites at and near the Idaho National Laboratory, Idaho, 1965–2009.—Continued

[**Local name:** Local well identifier used in this study. **Site identifier:** Unique numerical identifiers used to access well data (http://waterdata.usgs.gov/nwis). **Constituent:** Arsenic, barium, chromium, manganese, molybdenum, uranium, and zinc in micrograms per liter. Other constituents in milligrams per liter. P, phosphorus; N, nitrogen. **Trend:** +, increasing; –, decreasing]

Local name	Site identifier	Constituent	Start date	End date	Sample size	Mean	Median	Minimum	Maximum	Standard deviation	p value	Slope	Trend
					Wells—Continued								
IET 1 Disp	435153112420501	Chloride	01-01-1986	12-31-2009	21	21.4	20.0	16.6	30.0	4.29	0.840	0.024	no trend
		Sodium	01-01-1986	12-31-2009	20	18.1	18.2	14.9	23.0	2.37	0.332	-0.150	no trend
		Sulfate	01-01-1994	12-31-2009	14	34.3	31.7	25.0	66.0	9.62	0.000	0.690	+
		Nitrate + Nitrite as N	01-01-1986	12-31-2009	19	0.547	0.634	0.048	0.840	0.272	0.027	-0.032	–
		Orthophosphate as P	01-01-1986	12-31-2009	19	0.147	0.165	0.018	0.220	0.055	0.000	-0.007	–
		Ammonia	01-01-1986	12-31-2009	18	0.518	0.512	0.049	0.910	0.237	0.000	-0.044	–
Leo Rogers 1	432533112504901	Chloride	01-01-1980	12-31-2009	22	18.0	18.0	16.0	20.0	0.955	0.324	-0.030	no trend
		Sodium	01-01-1980	12-31-2009	21	16.8	17.0	14.0	22.0	1.42	0.728	0.002	no trend
No Name 1	435038112453401	Chloride	01-01-1990	12-31-2009	24	19.6	19.3	18.0	23.0	1.07	0.240	0.041	no trend
		Sodium	01-01-1990	12-31-2009	26	10.3	10.2	9.56	11.6	0.549	0.633	0.009	no trend
		Nitrate + Nitrite as N	01-01-1990	12-31-2009	27	0.641	0.642	0.550	0.806	0.053	0.110	-0.003	no trend
		Arsenic	01-01-1990	12-31-2009	23	1.90	2.00	1.40	2.50	0.268	0.165	-0.013	no trend
		Barium	01-01-1990	12-31-2009	23	68.4	68.0	60.8	73.1	2.88	0.095	0.257	no trend
		Chromium	01-01-1990	12-31-2009	23	8.48	8.45	6.57	13.0	1.38	0.005	-0.182	–
		Uranium	01-01-1990	12-31-2009	21	1.73	1.69	1.47	2.00	0.176	0.030	-0.025	–
NPR Test	433449112523101	Chloride	01-01-1986	12-31-2007	28	13.0	12.1	9.36	22.0	3.17	0.000	-0.434	–
		Sodium	01-01-1986	12-31-2007	25	7.58	7.57	6.98	8.50	0.478	0.000	-0.092	–
		Nitrate + Nitrite as N	01-01-1986	12-31-2007	25	0.707	0.621	0.523	1.40	0.210	0.000	-0.025	–
		Total organic carbon	01-01-1986	12-31-2007	13	1.62	0.700	0.201	7.28	2.06	0.434	-0.026	no trend
P and W 2	435419112453101	Chloride	01-01-1986	12-31-2009	43	10.2	7.40	5.00	66.1	9.65	0.327	-0.065	no trend
		Sodium	01-01-1986	12-31-2009	26	7.36	7.32	6.71	8.28	0.401	0.902	0.000	no trend
		Nitrate + Nitrite as N	01-01-1986	12-31-2009	25	0.429	0.393	0.318	0.88	0.117	0.154	0.008	no trend
PSTF Test	434941112454201	Chloride	01-01-1992	12-31-2009	24	6.48	6.34	5.90	9.02	0.607	0.314	0.013	no trend
		Sodium	01-01-1992	12-31-2009	24	6.85	6.78	6.00	7.81	0.397	0.871	-0.003	no trend
		Nitrate + Nitrite as N	01-01-1992	12-31-2009	26	0.589	0.592	0.142	0.781	0.105	0.471	-0.001	no trend
		Barium	01-01-1992	12-31-2009	23	67.0	67.0	61.4	71.0	2.25	0.078	-0.197	no trend
		Molybdenum	01-01-1992	12-31-2009	23	1.84	1.81	1.67	2.000	0.109	0.000	-0.020	–
		Uranium	01-01-1992	12-31-2009	22	1.62	1.58	1.00	2.00	0.234	0.245	-0.015	no trend
Site 4	433617112542001	Chloride	01-01-1980	12-31-2009	26	15.0	12.7	7.87	28.0	6.61	0.000	-0.920	–
		Sodium	01-01-1980	12-31-2009	18	8.64	8.70	7.03	11.0	1.27	0.000	-0.133	–
		Sulfate	01-01-1980	12-31-2009	12	21.3	20.2	19.1	29.0	2.92	0.651	-0.046	no trend
Site 9	433123112530101	Chloride	10-01-1990	12-31-2009	24	13.6	13.5	12.0	16.0	1.31	0.139	0.117	no trend
		Sodium	10-01-1990	12-31-2009	21	12.4	12.3	12.0	13.4	0.455	0.000	0.040	+
		Sulfate	10-01-1990	12-31-2009	17	23.6	24.0	21.0	25.2	1.22	0.000	0.134	+
		Nitrate + Nitrite as N	10-01-1990	12-31-2009	21	0.670	0.660	0.552	0.960	0.077	0.000	0.004	+
Site 14	434334112463101	Chloride	01-01-1980	12-31-2009	37	8.84	8.90	8.00	10.0	0.487	0.534	0.009	no trend
		Sodium	01-01-1980	12-31-2009	31	14.5	14.7	12.0	16.0	0.767	0.125	-0.023	no trend
		Nitrate + Nitrite as N	01-01-1980	12-31-2009	24	0.570	0.579	0.510	0.614	0.028	0.060	0.003	no trend

Table 3.   Statistical summaries and trend analyses of uncensored water-quality results for selected constituents in water from wells and surface-water sites at and near the Idaho National Laboratory, Idaho, 1965–2009.—Continued

[Local name: Local well identifier used in this study. Site identifier: Unique numerical identifiers used to access well data (http://waterdata.usgs.gov/nwis). Constituent: Arsenic, barium, chromium, manganese, molybdenum, uranium, and zinc in micrograms per liter. Other constituents in milligrams per liter. P, phosphorus; N, nitrogen. Trend: +, increasing; –, decreasing]

Local name	Site identifier	Constituent	Start date	End date	Sample size	Mean	Median	Minimum	Maximum	Standard deviation	p value	Slope	Trend
					Wells—Continued								
Site 17	43402711257570l	Chloride	01-01-1990	12-31-2009	20	10.9	10.4	9.30	15.0	1.46	0.189	-0.039	no trend
		Sodium	01-01-1990	12-31-2009	20	10.2	10.1	9.60	11.3	0.461	0.723	0.009	no trend
		Sulfate	01-01-1990	12-31-2009	15	20.1	20.2	16.0	21.8	1.40	0.050	0.152	no trend
		Nitrate + Nitrite as N	01-01-1990	12-31-2009	20	1.01	1.00	0.887	1.19	0.086	0.073	-0.007	no trend
Site 19	433522112582101	Chloride	01-01-1986	12-31-2009	33	11.8	11.0	9.68	16.0	1.79	0.002	-0.097	–
		Sodium	01-01-1986	12-31-2009	25	8.50	8.50	7.00	9.79	0.630	0.359	0.018	no trend
		Sulfate	01-01-1986	12-31-2009	16	23.1	21.0	19.4	35.4	4.82	0.788	0.030	no trend
SPERT 1	433252112520301	Chloride	01-01-1980	12-31-2009	43	23.0	21.0	12.0	43.6	7.55	0.003	0.322	+
		Sodium	01-01-1980	12-31-2009	30	13.3	12.5	8.00	23.5	3.20	0.002	0.201	+
		Nitrate + Nitrite as N	01-01-1980	12-31-2009	23	1.01	1.04	0.210	1.45	0.273	0.000	0.028	+
TRA 1	433521112573801	Chloride	01-01-1980	12-31-2009	40	11.1	11.0	8.00	15.0	1.49	0.012	-0.071	–
		Sodium	01-01-1980	12-31-2009	27	8.51	8.69	6.00	9.52	0.735	0.000	0.039	+
		Sulfate	01-01-1980	12-31-2009	14	20.7	20.8	18.4	24.1	1.62	0.030	0.160	+
TRA 3	433522112573501	Chloride	01-01-1980	12-31-2009	41	10.8	10.1	8.91	16.0	1.58	0.012	-0.069	–
		Sodium	01-01-1980	12-31-2009	29	8.78	8.91	5.00	11.0	1.30	0.035	0.044	+
		Sulfate	01-01-1980	12-31-2009	15	21.6	21.7	18.3	25.4	1.59	0.968	-0.000	no trend
TRA 4	433521112574201	Chloride	01-01-1980	12-31-2009	41	11.0	11.0	9.00	15.0	1.60	0.000	-0.092	–
		Sodium	01-01-1980	12-31-2009	29	7.53	7.89	5.00	8.58	0.915	0.000	0.056	+
		Sulfate	01-01-1980	12-31-2009	16	19.8	19.9	18.1	21.9	1.13	0.040	0.147	+
USGS 1	432700112470801	Chloride	01-01-1990	12-31-2009	26	12.9	13.0	11.0	14.0	0.654	0.022	0.060	+
		Sodium	01-01-1990	12-31-2009	26	14.8	15.0	13.7	16.2	0.610	0.457	0.013	no trend
		Nitrate + Nitrite as N	01-01-1990	12-31-2009	26	0.992	0.957	0.820	1.70	0.167	0.007	0.012	+
		Total organic carbon	01-01-1990	12-31-2009	14	0.710	0.454	0.200	2.03	0.563	0.093	0.017	no trend
USGS 2	433320112432301	Chloride	01-01-1990	12-31-2009	17	16.4	16.5	15.78	17.0	0.410	0.224	0.023	no trend
		Sodium	01-01-1990	12-31-2009	18	16.6	16.8	15.78	17.8	0.544	0.127	0.057	no trend
		Sulfate	01-01-1990	12-31-2009	15	14.6	14.4	13.00	17.2	1.32	0.000	0.229	+
		Nitrate + Nitrite as N	01-01-1990	12-31-2009	18	1.41	1.39	1.10	1.70	0.213	0.000	0.036	+
USGS 4	434657112282201	Chloride	01-01-1990	12-31-2009	26	35.8	35.64	28.3	51.0	5.10	0.000	-0.791	–
		Sodium	01-01-1990	12-31-2009	26	46.8	46.7	44.0	51.2	1.77	0.773	0.028	no trend
		Nitrate + Nitrite as N	01-01-1990	12-31-2009	26	4.44	4.57	1.70	5.00	0.595	0.234	-0.016	no trend
		Total organic carbon	01-01-1990	12-31-2009	15	1.92	1.36	0.805	4.70	1.22	0.110	-0.077	no trend
USGS 5	433543112493801	Chloride	01-01-1990	12-31-2009	25	8.95	8.98	7.10	10.4	0.745	0.002	0.096	+
		Sodium	01-01-1990	12-31-2009	24	7.59	7.54	6.99	8.34	0.363	0.083	0.035	no trend
		Nitrate + Nitrite as N	01-01-1990	12-31-2009	26	0.424	0.458	0.113	0.503	0.105	0.634	-0.001	no trend
USGS 6	434031112453701	Chloride	01-01-1990	12-31-2009	18	9.66	8.64	7.60	14.3	2.26	0.027	0.286	+
		Sodium	01-01-1990	12-31-2009	18	13.3	13.1	12.5	14.4	0.551	0.457	0.021	no trend
		Sulfate	01-01-1990	12-31-2009	15	18.4	18.1	17.3	20.5	0.985	0.230	0.100	no trend
		Nitrate + Nitrite as N	01-01-1990	12-31-2009	19	0.557	0.522	0.476	0.723	0.084	0.053	0.010	no trend

Table 3   43

Table 3. Statistical summaries and trend analyses of uncensored water-quality results for selected constituents in water from wells and surface-water sites at and near the Idaho National Laboratory, Idaho, 1965–2009.—Continued

[**Local name:** Local well identifier used in this study. **Site identifier:** Unique numerical identifiers used to access well data (http://waterdata.usgs.gov/nwis). **Constituent:** Arsenic, barium, chromium, manganese, molybdenum, uranium, and zinc in micrograms per liter. Other constituents in milligrams per liter. P, phosphorus; N, nitrogen. **Trend:** +, increasing; –, decreasing]

Local name	Site identifier	Constituent	Start date	End date	Sample size	Mean	Median	Minimum	Maximum	Standard deviation	p value	Slope	Trend
		Wells—Continued											
USGS 7	434915112443901	Chloride	01-01-1990	12-31-2009	26	9.11	8.93	8.60	12.0	0.655	0.820	0.002	no trend
		Sodium	01-01-1990	12-31-2009	26	23.2	23.0	21.9	25.0	0.870	0.940	0.000	no trend
		Nitrate + Nitrite as N	01-01-1990	12-31-2009	26	0.384	0.386	0.303	0.467	0.031	0.085	-0.002	no trend
		Arsenic	01-01-1990	12-31-2009	24	3.37	3.20	1.99	4.16	0.560	0.204	-0.046	no trend
		Barium	01-01-1990	12-31-2009	24	16.3	16.3	14.9	18.0	0.747	0.012	-0.092	–
		Manganese	01-01-1990	12-31-2009	24	2.88	2.65	1.88	5.06	0.901	0.357	0.039	no trend
		Uranium	01-01-1990	12-31-2009	22	2.10	2.11	1.91	2.32	0.117	0.684	-0.003	no trend
USGS 8	433121113115801	Chloride	01-01-1990	12-31-2009	32	8.16	7.85	6.92	11.0	0.933	0.000	-0.078	–
		Sodium	01-01-1990	12-31-2009	24	6.668	6.70	6.08	7.24	0.280	0.689	0.003	no trend
		Nitrate + Nitrite as N	01-01-1990	12-31-2009	24	0.875	0.90	0.650	0.941	0.062	0.139	0.003	no trend
USGS 9	432740113044501	Chloride	01-01-1987	12-31-2009	42	21.3	20.0	12.0	37.0	6.01	0.000	-0.777	–
		Sodium	01-01-1987	12-31-2009	33	12.3	12.1	7.00	19.0	1.99	0.000	-0.170	–
		Nitrate + Nitrite as N	01-01-1987	12-31-2009	25	0.698	0.657	0.373	1.90	0.260	0.653	-0.001	no trend
USGS 11	432336113064201	Chloride	09-01-1989	12-31-2009	38	11.6	11.1	8.55	16.0	1.99	0.000	-0.275	–
		Sodium	09-01-1989	12-31-2009	39	8.22	8.14	7.61	9.36	0.380	0.000	-0.030	–
		Nitrate + Nitrite as N	09-01-1989	12-31-2009	30	0.675	0.644	0.550	1.60	0.180	0.105	0.003	no trend
USGS 12	434126112550701	Chloride	01-01-1990	12-31-2009	44	31.2	33.5	12.4	42.0	8.32	0.018	-0.661	–
		Sodium	01-01-1990	12-31-2009	27	15.3	16.0	11.3	19.0	2.14	0.972	0.000	no trend
		Sulfate	01-01-1990	12-31-2009	43	31.9	33.0	22.3	38.0	4.41	0.000	-0.454	–
		Nitrate + Nitrite as N	01-01-1990	12-31-2009	47	1.76	1.89	0.460	2.20	0.411	0.656	0.000	no trend
		Bromide	01-01-1990	12-31-2009	23	0.080	0.080	0.060	0.100	0.011	0.007	0.004	+
USGS 14	432019112563201	Chloride	01-01-1989	12-31-2009	40	21.2	21.0	18.0	25.0	1.45	0.554	-0.015	no trend
		Sodium	01-01-1989	12-31-2009	8	16.1	17.0	12.0	17.0	1.73	0.781	0.000	no trend
USGS 15	434234112551701	Chloride	01-01-1990	12-31-2009	38	17.0	15.5	4.80	46.2	9.55	0.119	0.384	no trend
		Sodium	01-01-1990	12-31-2009	17	16.0	15.1	7.90	26.9	5.36	0.755	-0.183	no trend
		Sulfate	01-01-1990	12-31-2009	38	23.7	23.9	15.0	40.0	5.72	0.112	0.277	no trend
		Nitrate + Nitrite as N	01-01-1990	12-31-2009	37	1.06	0.991	0.260	3.19	0.699	0.013	0.044	+
		Bromide	01-01-1990	12-31-2009	25	0.037	0.040	0.020	0.080	0.016	0.860	0.000	no trend
USGS 17	433937112515401	Chloride	01-01-1989	12-31-2009	42	6.05	5.73	5.020	7.90	0.714	0.000	-0.078	–
		Sodium	01-01-1989	12-31-2009	24	7.05	7.06	6.30	7.73	0.343	0.035	0.022	+
		Sulfate	01-01-1989	12-31-2009	23	19.1	19.0	18.0	22.0	1.25	0.753	0.000	no trend
		Nitrate + Nitrite as N	01-01-1989	12-31-2009	42	0.416	0.352	0.250	2.00	0.319	0.003	0.004	+
		Fluoride	01-01-1989	12-31-2009	24	0.229	0.200	0.100	0.300	0.055	0.992	0.000	no trend
USGS 18	434540112440901	Chloride	01-01-1990	12-31-2009	18	10.3	10.2	8.94	12.0	0.784	0.017	0.092	+
		Sodium	01-01-1990	12-31-2009	18	12.1	12.0	11.57	13.3	0.428	0.204	0.018	no trend
		Sulfate	01-01-1990	12-31-2009	15	25.2	25.3	23.0	26.7	0.965	0.000	0.166	+
		Nitrate + Nitrite as N	01-01-1990	12-31-2009	19	0.616	0.613	0.573	0.750	0.040	0.641	0.000	no trend

Table 3. Statistical summaries and trend analyses of uncensored water-quality results for selected constituents in water from wells and surface-water sites at and near the Idaho National Laboratory, Idaho, 1965–2009.—Continued

[Local name: Local well identifier used in this study. Site identifier: Unique numerical identifiers used to access well data (http://waterdata.usgs.gov/nwis). Constituent: Arsenic, barium, chromium, manganese, molybdenum, uranium, and zinc in micrograms per liter. Other constituents in milligrams per liter. P, phosphorus; N, nitrogen. Trend: +, increasing; –, decreasing]

Local name	Site identifier	Constituent	Start date	End date	Sample size	Mean	Median	Minimum	Maximum	Standard deviation	p value	Slope	Trend
				Wells—Continued									
USGS 19	434426112575701	Chloride	01-01-1990	12-31-2009	31	10.6	10.1	9.04	15.0	1.52	0.279	-0.067	no trend
		Sodium	01-01-1990	12-31-2009	26	11.1	11.1	9.57	13.0	0.869	0.000	-0.145	–
		Nitrate + Nitrite as N	01-01-1990	12-31-2009	26	0.903	0.885	0.802	1.10	0.093	0.003	-0.007	–
USGS 22	433422113031701	Chloride	01-01-1990	12-31-2009	23	63.3	62.5	59.9	73.0	3.31	0.150	-0.127	no trend
		Sodium	01-01-1990	12-31-2009	19	22.3	22.2	20.7	24.0	0.787	0.828	0.000	no trend
USGS 23	434055112595901	Chloride	01-01-1990	12-31-2009	25	9.92	10.0	9.20	10.6	0.331	0.379	-0.010	no trend
		Sodium	01-01-1990	12-31-2009	26	9.39	9.21	8.28	10.8	0.713	0.037	0.083	+
		Nitrate + Nitrite as N	01-01-1990	12-31-2009	26	0.646	0.630	0.471	0.854	0.114	0.043	-0.010	–
USGS 26	435212112394001	Chloride	01-01-1990	12-31-2009	25	13.4	13.2	12.0	15.0	0.631	0.060	0.056	no trend
		Sodium	01-01-1990	12-31-2009	25	15.1	15.0	14.0	16.2	0.625	0.028	0.062	+
		Nitrate + Nitrite as N	01-01-1990	12-31-2009	26	0.812	0.842	0.210	0.966	0.132	0.015	0.005	+
		Arsenic	01-01-1990	12-31-2009	24	2.34	2.41	1.46	3.00	0.462	0.750	0.000	no trend
		Uranium	01-01-1990	12-31-2009	22	2.25	2.31	2.00	2.45	0.158	0.090	0.012	no trend
USGS 27	434851112321801	Chloride	01-01-1990	12-31-2009	32	59.6	58.9	50.0	76.0	5.37	0.000	-0.740	–
		Sodium	01-01-1990	12-31-2009	26	28.3	28.0	27.0	31.0	0.897	0.187	0.039	no trend
		Nitrate + Nitrite as N	01-01-1990	12-31-2009	26	2.48	2.49	2.30	2.72	0.100	0.050	0.010	no trend
		Total organic carbon	01-01-1990	12-31-2009	15	1.22	1.31	0.700	1.97	0.405	0.973	0.000	no trend
USGS 29	434407112285101	Chloride	01-01-1990	12-31-2009	20	28.2	26.7	22.0	53.1	6.28	0.798	0.020	no trend
		Sodium	01-01-1990	12-31-2009	19	21.1	20.1	14.5	42.2	5.37	0.676	-0.058	no trend
		Sulfate	01-01-1990	12-31-2009	15	18.6	16.2	15.0	40.3	6.88	0.659	0.037	no trend
		Nitrate + Nitrite as N	01-01-1990	12-31-2009	20	2.35	2.21	1.90	4.34	0.604	0.060	0.020	no trend
USGS 31	434625112342101	Chloride	01-01-1990	12-31-2009	20	24.7	24.0	19.0	29.3	3.35	0.000	0.587	+
		Sodium	01-01-1990	12-31-2009	20	15.8	15.8	15.0	17.4	0.654	0.097	0.045	no trend
		Sulfate	01-01-1990	12-31-2009	16	29.4	29.6	26.0	32.5	1.92	0.000	0.349	+
		Nitrate + Nitrite as N	01-01-1990	12-31-2009	20	0.919	0.899	0.810	1.06	0.081	0.000	0.014	+
USGS 32	434444112322101	Chloride	01-01-1990	12-31-2009	20	43.3	42.5	27.8	59.7	11.5	0.057	-0.707	no trend
		Sodium	01-01-1990	12-31-2009	20	18.5	18.0	16.9	21.1	1.27	0.134	-0.068	no trend
		Sulfate	01-01-1990	12-31-2009	16	37.4	37.2	28.0	45.0	5.28	0.067	-0.560	no trend
		Nitrate + Nitrite as N	01-01-1990	12-31-2009	19	1.38	1.38	1.02	1.78	0.254	0.209	-0.014	no trend
USGS 83	433023112561501	Chloride	08-01-1980	12-31-2009	49	10.0	10.7	4.00	17.0	2.61	0.290	0.076	no trend
		Sodium	08-01-1980	12-31-2009	34	9.37	9.79	6.00	10.7	1.04	0.142	0.031	no trend
		Nitrate + Nitrite as N	08-01-1980	12-31-2009	28	0.646	0.629	0.370	0.903	0.103	0.092	-0.007	no trend
USGS 86	432935113080001	Chloride	01-01-1987	12-31-2009	40	20.3	19.7	14.8	27.0	2.63	0.000	-0.331	–
		Sodium	01-01-1987	12-31-2009	25	11.3	11.3	10.1	12.1	0.558	0.257	-0.028	no trend
		Nitrate + Nitrite as N	01-01-1987	12-31-2009	24	1.40	1.45	0.758	1.70	0.199	0.045	-0.009	–

Table 3   45

Table 3. Statistical summaries and trend analyses of uncensored water-quality results for selected constituents in water from wells and surface-water sites at and near the Idaho National Laboratory, Idaho, 1965–2009.—Continued

[**Local name:** Local well identifier used in this study. **Site identifier:** Unique numerical identifiers used to access well data (http://waterdata.usgs.gov/nwis). **Constituent:** Arsenic, barium, chromium, manganese, molybdenum, uranium, and zinc in micrograms per liter. Other constituents in milligrams per liter. P, phosphorus; N, nitrogen. **Trend:** +, increasing; −, decreasing]

Local name	Site identifier	Constituent	Start date	End date	Sample size	Mean	Median	Minimum	Maximum	Standard deviation	p value	Slope	Trend
				Wells—Continued									
USGS 97	433807112551501	Chloride	01-01-1986	12-31-2009	69	32.5	33.0	25.0	39.0	3.15	0.000	0.358	+
		Sodium	01-01-1986	12-31-2009	32	15.5	15.2	13.0	17.5	1.04	0.000	0.123	+
		Sulfate	01-01-1986	12-31-2009	50	34.4	34.3	27.0	39.0	2.21	0.002	−0.247	−
		Nitrate + Nitrite as N	01-01-1986	12-31-2009	52	1.96	2.00	0.330	2.46	0.295	0.000	0.016	+
		Bromide	01-01-1986	12-31-2009	27	0.078	0.080	0.060	0.090	0.010	0.042	0.003	+
		Arsenic	01-01-1986	12-31-2009	25	1.79	1.84	1.10	2.29	0.319	0.144	−0.015	no trend
		Barium	01-01-1986	12-31-2009	25	137	136	122	200	14.6	0.025	−0.670	−
		Molybdenum	01-01-1986	12-31-2009	23	1.30	1.31	1.00	1.60	0.211	0.000	0.045	+
		Uranium	01-01-1986	12-31-2009	22	2.29	2.27	2.00	3.00	0.245	0.938	−0.001	no trend
USGS 98	433657112563601	Chloride	01-01-1986	12-31-2009	66	15.2	14.4	13.0	24.0	2.08	0.012	−0.114	−
		Sodium	01-01-1986	12-31-2009	30	10.1	9.89	9.05	13.0	0.916	0.920	0.000	no trend
		Sulfate	01-01-1986	12-31-2009	44	21.5	21.3	19.0	25.0	1.19	0.666	0.000	no trend
		Nitrate + Nitrite as N	01-01-1986	12-31-2009	48	1.14	1.10	0.938	3.00	0.284	0.821	0.000	no trend
		Bromide	01-01-1986	12-31-2009	24	0.042	0.040	0.030	0.050	0.006	0.932	0.000	no trend
		Barium	01-01-1986	12-31-2009	26	51.9	52.0	38.2	100	13.6	0.399	0.394	no trend
		Uranium	01-01-1986	12-31-2009	22	1.59	1.56	1.00	2.00	0.277	0.523	0.013	no trend
		Zinc	01-01-1986	12-31-2009	23	134	151	2.12	215	72.4	0.000	−12.9	−
USGS 99	433705112552101	Chloride	01-01-1986	12-31-2009	68	20.6	21.0	14.0	25.0	2.24	0.000	0.193	+
		Sulfate	01-01-1986	12-31-2009	42	26.6	27.0	23.0	30.0	1.37	0.290	0.023	no trend
		Sodium	01-01-1986	12-31-2009	25	13.8	14.0	8.00	16.1	1.74	0.000	0.163	+
		Nitrate + Nitrite as N	01-01-1986	12-31-2009	29	1.56	1.60	1.40	1.70	0.068	0.900	0.000	no trend
		Bromide	01-01-1986	12-31-2009	26	0.055	0.050	0.040	0.070	0.008	0.422	0.002	no trend
		Total organic carbon	01-01-1986	12-31-2009	30	0.750	0.450	0.100	3.80	0.820	0.018	0.092	+
USGS 100	433503112400701	Chloride	01-01-1986	12-31-2009	49	16.24	16.0	12.0	23.0	1.81	0.080	0.049	+
		Sodium	01-01-1986	12-31-2009	23	16.1	16.4	10.0	18.0	1.68	0.005	0.072	+
		Sulfate	01-01-1986	12-31-2009	16	17.8	17.0	11.0	27.6	4.18	0.023	0.302	+
USGS 101	433255112381801	Chloride	01-01-1986	12-31-2009	40	8.58	8.63	5.80	11.0	1.07	0.000	0.070	+
		Sodium	01-01-1986	12-31-2009	31	14.0	14.1	10.0	15.0	0.985	0.195	0.028	no trend
		Nitrate + Nitrite as N	01-01-1986	12-31-2009	24	0.933	0.916	0.720	1.15	0.122	0.000	0.022	+
USGS 102	433853112551601	Chloride	01-01-1990	12-31-2009	42	31.4	31.0	23.0	36.2	3.05	0.955	0.000	no trend
		Sodium	01-01-1990	12-31-2009	19	15.2	15.1	13.0	17.0	0.990	0.000	0.129	+
		Sulfate	01-01-1990	12-31-2009	40	32.9	33.9	23.0	39.0	3.10	0.072	−0.214	no trend
		Bromide	01-01-1990	12-31-2009	25	0.078	0.080	0.050	0.120	0.014	0.035	0.003	+
		Nitrate + Nitrite as N	01-01-1990	12-31-2009	40	1.83	1.90	0.300	2.10	0.295	0.002	0.018	+
		Total organic carbon	01-01-1990	12-31-2009	29	0.738	0.400	0.200	4.60	0.855	0.028	0.063	+
USGS 103	432714112560701	Chloride	01-01-1986	12-31-2005	63	15.4	15.9	11.0	20.0	1.58	0.117	0.054	no trend
		Sodium	01-01-1986	12-31-2005	29	13.0	13.2	8.00	15.0	1.44	0.927	0.000	no trend
		Nitrate + Nitrite as N	01-01-1986	12-31-2005	21	0.734	0.748	0.510	0.875	0.075	0.003	0.015	+

Table 3. Statistical summaries and trend analyses of uncensored water-quality results for selected constituents in water from wells and surface-water sites at and near the Idaho National Laboratory, Idaho, 1965–2009.—Continued

[**Local name:** Local well identifier used in this study. **Site identifier:** Unique numerical identifiers used to access well data (http://waterdata.usgs.gov/nwis). **Constituent:** Arsenic, barium, chromium, manganese, molybdenum, uranium, and zinc in micrograms per liter. Other constituents in milligrams per liter. P, phosphorus; N, nitrogen. **Trend:** +, increasing; –, decreasing]

Local name	Site identifier	Constituent	Start date	End date	Sample size	Mean	Median	Minimum	Maximum	Standard deviation	p value	Slope	Trend
					Wells—Continued								
USGS 107	432942112532801	Chloride	01-01-1983	12-31-2009	47	20.1	21.0	14.0	25.0	2.52	0.000	0.199	+
		Sodium	01-01-1983	12-31-2009	31	17.1	17.5	10.0	19.1	1.56	0.002	0.105	+
		Nitrate + Nitrite as N	01-01-1983	12-31-2009	30	1.10	1.08	0.880	2.50	0.281	0.000	0.015	+
USGS 109	432701113025601	Chloride	07-01-1987	12-31-2009	40	15.3	14.6	12.1	22.0	2.75	0.000	-0.256	-
		Sodium	07-01-1987	12-31-2009	32	11.9	12.0	10.4	14.1	0.894	0.644	0.000	no trend
		Nitrate + Nitrite as N	07-01-1987	12-31-2009	24	0.602	0.605	0.418	0.679	0.059	0.000	0.007	+
USGS 110	432717112501501	Chloride	01-01-1983	12-31-1992	27	18.2	18.0	13.0	25.0	3.15	0.000	0.922	+
		Sodium	01-01-1983	12-31-1992	11	15.6	16.0	9.00	19.0	2.50	0.053	0.455	no trend
USGS 110A	432717112501502	Chloride	01-01-1995	12-31-2009	14	19.4	19.1	18.1	22.0	1.00	0.120	0.031	no trend
		Sodium	01-01-1995	12-31-2009	18	17.3	17.2	16.4	18.1	0.590	0.571	-0.017	no trend
		Nitrate + Nitrite as N	01-01-1995	12-31-2009	19	1.15	1.25	0.035	1.34	0.311	0.346	0.005	no trend
USGS 117	432955113025901	Chloride	01-01-1987	12-31-2009	61	13.3	13.0	10.4	16.0	1.18	0.000	-0.140	-
		Sodium	01-01-1987	12-31-2009	24	10.1	10.1	7.00	11.5	0.923	0.022	0.053	+
		Sulfate	01-01-1987	12-31-2009	18	18.0	17.7	16.0	21.0	1.19	0.965	0.008	no trend
		Nitrate + Nitrite as N	01-01-1987	12-31-2009	24	0.682	0.677	0.615	0.780	0.041	0.002	-0.003	-
USGS 119	432945113023401	Chloride	01-01-1987	12-31-2009	65	9.66	9.31	8.0	14.0	1.17	0.050	-0.058	no trend
		Sodium	01-01-1987	12-31-2009	26	10.8	11.0	9.97	14.0	0.787	0.853	0.000	no trend
		Sulfate	01-01-1987	12-31-2009	17	34.1	33.8	28.0	42.5	4.45	0.000	0.863	+
		Nitrate + Nitrite as N	01-01-1987	12-31-2009	25	1.40	1.40	1.20	1.80	0.123	0.526	0.001	no trend
USGS 121	433450112560301	Chloride	01-01-1991	12-31-2009	32	12.9	12.6	9.62	21.0	2.23	0.010	-0.207	-
		Sodium	01-01-1991	12-31-2009	20	7.45	7.47	6.04	8.50	0.486	0.376	-0.017	no trend
		Sulfate	01-01-1991	12-31-2009	18	22.7	22.9	21.1	24.0	0.770	0.705	-0.017	no trend
		Nitrate + Nitrite as N	01-01-1991	12-31-2009	21	0.788	0.764	0.711	1.10	0.088	0.254	-0.002	no trend
USGS 125	432602113052801	Chloride	01-01-1995	12-31-2009	25	12.9	13.3	10.4	15.0	1.23	0.000	-0.232	-
		Sodium	01-01-1995	12-31-2009	23	11.8	11.8	11.0	13.0	0.529	0.037	-0.046	-
		Nitrate + Nitrite as N	01-01-1995	12-31-2009	23	0.566	0.570	0.490	0.623	0.029	0.346	0.001	no trend
USGS 126A	435529112471301	Chloride	01-01-2000	12-31-2009	12	7.98	7.99	7.22	8.53	0.317	0.391	0.032	no trend
		Sodium	01-01-2000	12-31-2009	12	8.70	8.63	8.00	9.66	0.458	0.561	-0.046	no trend
		Nitrate + Nitrite as N	01-01-2000	12-31-2009	11	0.518	0.523	0.479	0.533	0.015	0.902	-0.000	no trend
USGS 126B	435529112471401	Chloride	01-01-2000	12-31-2009	10	8.02	8.04	7.33	8.56	0.343	0.621	0.017	no trend
		Sodium	01-01-2000	12-31-2009	10	8.61	8.47	7.99	9.27	0.393	0.210	-0.025	no trend
		Nitrate + Nitrite as N	01-01-2000	12-31-2009	11	0.590	0.524	0.483	1.30	0.236	0.688	-0.001	no trend
WS INEL 1	433716112563601	Chloride	01-01-1986	12-31-2009	54	83.0	81.0	32.9	145	30.6	0.000	-5.44	-
		Sodium	01-01-1986	12-31-2009	23	15.7	16.0	11.4	22.0	3.22	0.000	-0.324	-
		Sulfate	01-01-1986	12-31-2009	38	44.0	41.7	30.0	61.0	8.79	0.000	-1.46	-
		Nitrate + Nitrite as N	01-01-1986	12-31-2009	25	4.50	4.60	1.40	5.60	0.985	0.000	-0.383	-
		Total organic carbon	01-01-1986	12-31-2009	26	0.877	0.800	0.500	3.50	0.562	0.686	0.000	no trend
		Bromide	01-01-1986	12-31-2009	26	0.279	0.290	0.190	0.350	0.047	0.000	-0.023	-

Table 3   47

Table 3. Statistical summaries and trend analyses of uncensored water-quality results for selected constituents in water from wells and surface-water sites at and near the Idaho National Laboratory, Idaho, 1965–2009.—Continued

[**Local name:** Local well identifier used in this study. **Site identifier:** Unique numerical identifiers used to access well data (http://waterdata.usgs.gov/nwis). **Constituent:** Arsenic, barium, chromium, manganese, molybdenum, uranium, and zinc in micrograms per liter. Other constituents in milligrams per liter. P, phosphorus; N, nitrogen. **Trend:** +, increasing; –, decreasing]

Local name	Site identifier	Constituent	Start date	End date	Sample size	Mean	Median	Minimum	Maximum	Standard deviation	p value	Slope	Trend
				Surface-water sites									
Big Lost River at Experimental Dairy Farm, near Howe	13132545	Chloride	01-01-1981	12-31-2009	25	4.66	5.00	2.13	7.00	1.17	0.072	-0.031	no trend
Big Lost River below INL diversion, near Arco	13132520	Chloride	01-01-1981	12-31-2009	29	4.97	4.97	2.06	9.00	1.68	0.000	-0.122	–
Big Lost River below Mackay Reservoir, near Mackay	13127700	Chloride	01-01-1985	12-31-2009	34	3.33	3.04	2.29	5.00	0.715	0.000	-0.085	–
Big Lost River near Arco	13132500	Chloride	01-01-1965	12-31-2009	90	6.02	5.75	2.09	14.0	2.20	0.140	-0.025	no trend
		Sodium	01-01-1965	12-31-2009	58	9.38	9.25	6.20	14.0	1.60	0.297	-0.043	no trend
		Sulfate	01-01-1965	12-31-2009	58	21.6	22.0	9.60	29.0	3.72	0.808	0.000	no trend
		Calcium	01-01-1965	12-31-2009	58	58.4	58.5	35.0	74.0	8.01	0.511	0.105	no trend
		Magnesium	01-01-1965	12-31-2009	58	15.0	15.0	10.0	19.0	1.78	0.811	0.000	no trend
		Silica	01-01-1965	12-31-2009	58	13.7	14.0	9.60	19.0	1.76	0.644	0.000	no trend
		Nitrate + Nitrite as N	01-01-1965	12-31-2009	23	0.283	0.230	0.020	2.00	0.395	0.244	-0.012	no trend
		Fluoride	01-01-1965	12-31-2009	58	0.298	0.300	0.200	0.600	0.066	1.00	0.000	no trend
		Potassium	01-01-1965	12-31-2009	58	1.77	1.70	1.20	2.80	0.346	0.780	0.000	no trend
Birch Creek at Blue Dome, near Reno	13117020	Chloride	01-01-1970	12-31-2009	50	5.18	4.70	3.69	11.0	1.30	0.002	-0.033	–
Little Lost River near Howe	13119000	Chloride	01-01-1965	12-31-2009	86	9.49	9.03	1.00	19.7	3.65	0.030	0.081	+
		Sodium	01-01-1965	12-31-2009	33	8.13	7.30	4.20	15.0	2.70	0.382	-0.091	no trend
		Sulfate	01-01-1965	12-31-2009	32	16.5	16.5	8.90	31.0	5.23	0.664	-0.045	no trend
		Calcium	01-01-1965	12-31-2009	33	37.7	39.0	20.0	52.0	7.27	0.497	-0.202	no trend
		Magnesium	01-01-1965	12-31-2009	33	14.9	15.0	9.20	21.0	2.97	0.648	0.000	no trend
		Silica	01-01-1965	12-31-2009	32	13.2	13.0	10.0	16.0	1.48	0.301	-0.108	no trend
		Nitrate + Nitrite as N	01-01-1965	12-31-2009	23	0.721	0.180	0.030	7.50	1.70	0.634	-0.005	no trend
		Potassium	01-01-1965	12-31-2009	31	1.41	1.30	1.00	2.50	0.391	0.813	0.000	no trend
Mud Lake near Terreton	13115000	Chloride	01-01-1965	12-31-2009	76	8.07	8.00	2.00	18.2	2.30	0.000	0.059	+
		Sodium	01-01-1965	12-31-2009	23	10.3	10.0	7.60	16.0	1.78	0.823	0.000	no trend
		Sulfate	01-01-1965	12-31-2009	24	7.79	7.45	4.10	17.0	2.50	0.995	-0.020	no trend
		Calcium	01-01-1965	12-31-2009	23	29.5	31.0	15.0	47.0	6.68	0.104	-0.682	no trend
		Magnesium	01-01-1965	12-31-2009	23	8.81	8.60	7.30	13.0	1.13	0.472	0.023	no trend
		Silica	01-01-1965	12-31-2009	23	23.9	26.0	7.70	31.0	6.30	0.070	-0.434	no trend
		Nitrate + Nitrite as N	01-01-1965	12-31-2009	15	0.382	0.250	0.030	1.30	0.361	0.065	-0.035	no trend
		Fluoride	01-01-1965	12-31-2009	24	0.413	0.400	0.200	0.500	0.095	0.908	0.000	no trend
		Potassium	01-01-1965	12-31-2009	20	2.46	2.40	1.30	4.40	0.636	0.686	-0.008	no trend

Table 4. Statistical summaries and trend analyses of censored water-quality results for selected constituents in water from wells and surface-water sites at and near the Idaho National Laboratory, Idaho, 1980–2009.

[**Local name:** Local well identifier used in this study. **Site identifier:** Unique numerical identifiers used to access well data (http://waterdata.usgs.gov/nwis). **Constituent**: Aluminum, antimony, arsenic, barium, cobalt, chromium, copper, manganese, molybdenum, nickel, lead, thallium, and zinc in micrograms per liter. Other constituents in milligrams per liter. **Trend:** +, increasing; –, decreasing. NA, not applicable]

Local name	Site identifier	Constituent	Start date	End date	Sample size	Number of censored values	Mean	Median	Minimum	Maximum	Standard deviation	p value	Slope	Trend
				Wells										
ANP 9	434856112400001	Aluminum	01-01-1994	12-31-2009	23	5	2.76	3.00	1.00	9.00	2.009	0.068	-0.241	no trend
		Antimony	01-01-1994	12-31-2009	23	14	0.119	0.120	0.102	1.00	0.018	0.771	-0.001	no trend
		Cobalt	01-01-1994	12-31-2009	23	15	0.061	0.067	0.012	1.00	0.058	0.584	-0.008	no trend
		Copper	01-01-1994	12-31-2009	23	17	0.354	0.334	0.238	1.00	0.192	0.970	-0.016	no trend
		Manganese	01-01-1994	12-31-2009	23	10	1.71	0.752	0.200	12.6	2.811	0.484	0.020	no trend
		Orthophosphate	01-01-1994	12-31-2009	24	10	0.011	0.009	0.009	0.021	0.004	0.110	0.000	no trend
		Total organic carbon	01-01-1994	12-31-2009	16	7	0.754	0.318	0.100	6.00	1.472	0.604	-0.001	no trend
		Lead	01-01-1994	12-31-2009	14	12	0.113	NA	0.030	1.00	0.010	0.687	0.013	no trend
ARBOR Test	433509112384801	Chromium	01-01-1989	12-31-2009	28	18	1.70	1.50	1.00	14.0	1.076	0.252	0.101	no trend
AREA 2	433223112470201	Orthophosphate	01-01-1992	12-31-2009	18	7	0.012	0.010	0.009	0.024	0.005	1.00	-0.000	no trend
Badging Facility Well	433042112535101	Orthophosphate	01-01-1985	12-31-2009	22	1	0.016	0.014	0.010	0.030	0.006	0.424	0.000	no trend
		Orthophosphate	01-01-1983	12-31-2009	21	2	0.018	0.020	0.010	0.029	0.005	0.274	0.000	no trend
		Chromium	01-01-1983	12-31-2009	38	9	6.14	6.00	3.00	50.0	1.870	0.100	0.115	no trend
EBR 1	433051113002601	Orthophosphate	01-01-1980	12-31-2009	15	7	0.012	0.012	0.009	0.020	0.003	0.830	0.000	no trend
		Chromium	01-01-1980	12-31-2009	26	6	6.96	6.79	5.00	14.0	1.394	0.928	0.009	no trend
		Total organic carbon	01-01-1980	12-31-2009	15	5	0.456	0.255	0.100	1.34	0.431	0.837	0.005	no trend
Fire Station 2	433548112562301	Chromium	01-01-1980	12-31-1996	37	22	5.72	4.00	1.00	50.0	3.951	0.633	0.201	no trend
Highway 3	433256113002501	Orthophosphate	01-01-1980	12-31-2009	29	1	0.025	0.013	0.009	0.290	0.051	0.955	0.000	no trend
		Chromium	01-01-1980	12-31-2009	24	16	1.80	1.57	1.00	14.0	1.145	0.724	-0.018	no trend
		Total organic carbon	01-01-1980	12-31-2009	15	3	0.599	0.390	0.100	2.14	0.642	0.539	0.017	no trend
No Name 1	435038112453401	Aluminum	01-01-1990	12-31-2009	23	3	3.13	1.82	1.02	12.0	2.524	0.027	-0.301	–
		Antimony	01-01-1990	12-31-2009	22	14	0.199	0.171	0.142	1.00	0.122	0.887	-0.003	no trend
		Cobalt	01-01-1990	12-31-2009	23	14	0.085	0.085	0.040	3.00	0.049	1.00	0.001	no trend
		Copper	01-01-1990	12-31-2009	23	15	0.334	0.256	0.218	10.0	0.220	0.973	0.006	no trend
		Nickel	01-01-1990	12-31-2009	23	14	0.619	0.232	0.050	10.0	0.825	0.080	0.174	no trend
		Thallium	01-01-1990	12-31-2009	22	21	0.045	NA	0.040	0.500	NA	0.799	-0.042	no trend
		Zinc	01-01-1990	12-31-2009	23	12	1.59	0.877	0.600	6.00	1.481	0.015	-0.310	–
		Manganese	01-01-1990	12-31-2009	23	2	1.60	1.52	0.628	3.27	0.780	0.614	0.031	no trend
		Orthophosphate	01-01-1990	12-31-2009	27	2	0.018	0.019	0.010	0.029	0.004	0.983	-0.000	no trend
		Total organic carbon	01-01-1990	12-31-2009	16	3	1.36	0.511	0.221	5.77	1.641	1.00	-0.004	no trend
		Molybdenum	01-01-1990	12-31-2009	23	1	5.98	6.00	5.65	10.0	0.264	1.00	-0.000	no trend
NPR Test	433449112523101	Orthophosphate	01-01-1986	12-31-2009	27	2	0.019	0.020	0.005	0.030	0.007	0.071	-0.000	no trend
		Chromium	01-01-1986	12-31-2009	26	6	7.05	6.90	1.20	14.0	2.545	0.185	-0.120	no trend
		Total organic carbon	01-01-1986	12-31-2009	15	2	1.50	0.701	0.201	7.28	1.953	0.803	-0.009	no trend
P and W 2	435419112453101	Orthophosphate	01-01-1986	12-31-2009	25	6	0.013	0.012	0.010	0.020	0.003	0.166	0.000	no trend
		Chromium	01-01-1986	12-31-2009	28	17	1.99	1.82	0.749	14.0	1.509	0.798	-0.039	no trend

Table 4   49

Table 4. Statistical summaries and trend analyses of censored water-quality results for selected constituents in water from wells and surface-water sites at and near the Idaho National Laboratory, Idaho, 1980–2009.—Continued

[**Local name:** Local well identifier used in this study. **Site identifier:** Unique numerical identifiers used to access well data (http://waterdata.usgs.gov/nwis). **Constituent:** Aluminum, antimony, arsenic, barium, cobalt, chromium, copper, manganese, molybdenum, nickel, lead, thallium, and zinc in micrograms per liter. Other constituents in milligrams per liter. **Trend:** +, increasing; –, decreasing. NA, not applicable]

Local name	Site identifier	Constituent	Start date	End date	Sample size	Number of censored values	Mean	Median	Minimum	Maximum	Standard deviation	p value	Slope	Trend
					Wells—Continued									
PSTF Test	434941112454201	Aluminum	01-01-1990	12-31-2009	23	2	4.25	3.00	1.09	29.04	5.71	0.352	-0.133	no trend
		Antimony	01-01-1990	12-31-2009	23	14	0.128	0.130	0.107	1.00	0.033	0.698	-0.002	no trend
		Arsenic	01-01-1990	12-31-2009	24	2	1.87	1.99	1.00	2.79	0.391	0.449	-0.001	--
		Chromium	01-01-1990	12-31-2009	24	1	3.54	3.00	1.00	9.00	1.43	0.001	-0.130	--
		Cobalt	01-01-1990	12-31-2009	23	15	0.051	0.041	0.021	1.00	0.038	0.584	-0.006	no trend
		Copper	01-01-1990	12-31-2009	23	17	0.233	0.231	0.141	1.00	0.167	0.911	0.025	no trend
		Lead	01-01-1990	12-31-2009	24	22	0.042	0.031	0.031	1.00	0.078	0.900	0.027	no trend
		Manganese	01-01-1990	12-31-2009	23	13	0.252	0.109	0.050	2.80	0.591	0.214	0.039	no trend
		Nickel	01-01-1990	12-31-2009	23	13	0.376	0.265	0.060	1.19	0.425	0.951	-0.000	no trend
		Thallium	01-01-1990	12-31-2009	23	22	0.501	NA	0.020	0.501	NA	0.921	0.000	no trend
		Zinc	01-01-1990	12-31-2009	23	1	2.36	2.00	1.00	5.89	1.29	0.398	-0.055	no trend
		Orthophosphate	01-01-1990	12-31-2009	27	2	0.019	0.020	0.010	0.031	0.005	0.237	0.000	no trend
		Total organic carbon	01-01-1980	12-31-2009	16	3	0.550	0.234	0.100	1.76	0.559	0.177	0.019	no trend
Site 4	433617112542001	Chromium	01-01-1980	12-31-2009	24	1	8.18	7.96	5.00	11.49	1.48	0.534	-0.032	no trend
Site 9	433123112530101	Orthophosphate	10-01-1990	12-31-2009	21	5	0.015	0.013	0.010	0.040	0.007	0.326	0.000	no trend
Site 14	434343112463101	Orthophosphate	01-01-1980	12-31-2009	24	8	0.014	0.010	0.010	0.050	0.008	0.052	0.000	no trend
		Chromium	01-01-1980	12-31-2009	31	10	4.62	5.00	0.000	14.00	2.12	0.231	0.064	no trend
		Total organic carbon	01-01-1980	12-31-2009	14	4	0.492	0.549	0.100	1.00	0.363	0.336	-0.037	no trend
Site 17	434027112575701	Orthophosphate	01-01-1990	12-31-2009	20	1	0.016	0.016	0.010	0.031	0.005	0.974	-0.000	no trend
Site 19	433522112582101	Chromium	01-01-1986	12-31-2009	27	8	3.76	3.66	1.00	50.0	2.28	0.388	0.074	no trend
SPERT 1	433525112520301	Orthophosphate	01-01-1980	12-31-2009	23	2	0.017	0.018	0.010	0.032	0.006	0.596	0.000	no trend
TRA 1	433521112573801	Chromium	01-01-1980	12-31-2009	31	17	3.03	3.00	1.00	50.0	1.91	0.688	0.017	no trend
TRA 3	433522112573501	Chromium	01-01-1980	12-31-2009	32	15	3.33	3.14	0.000	50.0	1.85	0.679	0.013	no trend
TRA 4	433521112574201	Chromium	01-01-1980	12-31-2009	31	15	3.57	3.67	2.00	50.0	1.44	0.665	0.039	no trend
USGS 1	432700112470801	Orthophosphate	01-01-1990	12-31-2009	26	18	0.011	0.009	0.009	0.020	0.004	0.140	0.001	no trend
		Chromium	01-01-1990	12-31-2009	24	14	2.33	2.00	1.09	14.0	1.20	0.807	-0.045	no trend
USGS 2	433320112432301	Orthophosphate	01-01-1990	12-31-2009	18	8	0.014	0.010	0.010	0.028	0.006	0.501	0.000	no trend
USGS 4	434657112282201	Orthophosphate	01-01-1990	12-31-2009	26	2	0.017	0.018	0.010	0.027	0.005	0.092	0.000	no trend
		Chromium	01-01-1990	12-31-2009	24	2	11.4	11.0	8.15	14.2	1.86	0.056	-0.185	no trend
USGS 5	433543112493801	Orthophosphate	01-01-1990	12-31-2009	26	10	0.011	0.011	0.004	0.090	0.006	0.253	0.000	no trend
		Total organic carbon	01-01-1990	12-31-2009	15	1	2.15	1.02	0.350	12.6	3.16	0.519	-0.047	no trend
USGS 6	434031112453701	Orthophosphate	01-01-1990	12-31-2009	19	1	0.015	0.013	0.010	0.023	0.005	0.015	0.000	+

Table 4. Statistical summaries and trend analyses of censored water-quality results for selected constituents in water from wells and surface-water sites at and near the Idaho National Laboratory, Idaho, 1980–2009.—Continued

[**Local name:** Local well identifier used in this study. **Site identifier:** Unique numerical identifiers used to access well data (http://waterdata.usgs.gov/nwis). **Constituent:** Aluminum, antimony, arsenic, barium, cobalt, chromium, copper, manganese, molybdenum, nickel, lead, thallium, and zinc in micrograms per liter. Other constituents in milligrams per liter. **Trend:** +, increasing; –, decreasing. NA, not applicable]

Local name	Site identifier	Constituent	Start date	End date	Sample size	Number of censored values	Mean	Median	Minimum	Maximum	Standard deviation	p value	Slope	Trend
				Wells—Continued										
USGS 7	434915112443901	Aluminum	01-01-1990	12-31-2009	23	2	3.24	3.00	1.510	13.4	1.58	0.021	-0.236	–
		Antimony	01-01-1990	12-31-2009	23	15	0.115	0.107	0.088	1.00	0.042	0.974	-0.001	no trend
		Chromium	01-01-1990	12-31-2009	24	1	2.68	2.56	1.580	5.00	0.830	0.000	-0.117	–
		Cobalt	01-01-1990	12-31-2009	24	15	0.045	0.043	0.013	3.00	0.029	0.953	-0.001	no trend
		Copper	01-01-1990	12-31-2009	24	19	0.248	0.225	0.129	10.0	0.210	0.676	-0.064	no trend
		Molybdenum	01-01-1990	12-31-2009	24	1	3.83	3.91	3.000	10.0	0.243	0.581	-0.003	no trend
		Nickel	01-01-1990	12-31-2009	24	11	0.497	0.384	0.116	10.0	0.420	0.299	-0.011	no trend
		Thallium	01-01-1990	12-31-2009	22	20	0.063	0.033	0.033	0.690	0.194	0.944	-0.077	no trend
		Zinc	01-01-1990	12-31-2009	24	14	2.30	0.841	0.600	29.0	5.91	0.116	-0.094	no trend
		Orthophosphate	01-01-1990	12-31-2009	26	22	0.011	0.010	0.010	0.020	0.003	0.360	0.001	no trend
		Total organic carbon	01-01-1990	12-31-2009	13	4	0.636	0.319	0.100	3.16	0.918	1.00	0.002	no trend
USGS 8	433121113115801	Orthophosphate	01-01-1990	12-31-2009	24	7	0.013	0.011	0.010	0.020	0.003	0.187	0.000	no trend
		Chromium	01-01-1990	12-31-2009	24	14	3.16	2.73	.881	14.0	1.70	1.00	-0.016	no trend
USGS 9	432740113044501	Total organic carbon	01-01-1990	12-31-2009	15	3	0.873	0.500	.100	3.42	1.04	0.801	0.009	no trend
		Orthophosphate	01-01-1987	12-31-2009	25	3	0.017	0.018	0.010	0.024	0.005	0.589	0.000	no trend
		Chromium	01-01-1987	12-31-2009	27	12	4.85	4.30	2.408	14.0	2.44	0.947	-0.000	no trend
		Total organic carbon	01-01-1987	12-31-2009	16	3	0.953	0.600	0.213	3.32	0.946	0.523	-0.020	no trend
USGS 11	432336113064201	Orthophosphate	09-01-1989	12-31-2009	30	10	0.012	0.011	0.010	0.180	0.003	0.034	0.000	+
		Chromium	09-01-1989	12-31-2009	30	13	4.36	3.41	1.100	15.38	2.88	0.587	0.075	no trend
USGS 12	434126112550701	Total organic carbon	09-01-1989	12-31-2009	16	3	0.797	0.379	0.100	3.50	0.953	0.582	0.011	no trend
		Orthophosphate	01-01-1990	12-31-2009	40	3	0.022	0.020	0.010	0.090	0.007	0.004	0.000	+
		Chromium	01-01-1990	12-31-2009	27	5	7.32	7.00	5.000	14.0	1.91	1.00	0.002	no trend
		Total organic carbon	01-01-1990	12-31-2009	42	1	0.958	0.400	0.228	7.68	1.38	0.026	0.028	+
USGS 15	434234112551701	Fluoride	01-01-1990	12-31-2009	25	1	0.196	0.200	0.130	0.400	0.061	0.637	0.000	no trend
		Orthophosphate	01-01-1990	12-31-2009	35	7	0.018	0.020	0.010	0.040	0.007	0.072	0.000	no trend
		Total organic carbon	01-01-1990	12-31-2009	25	3	0.836	0.300	0.100	3.80	1.16	0.072	0.060	no trend
		Fluoride	01-01-1990	12-31-2009	25	1	0.164	0.100	0.100	0.600	0.119	0.011	-0.000	–
USGS 17	433937112515401	Orthophosphate	01-01-1989	12-31-2009	37	7	0.015	0.014	0.010	0.036	0.005	0.332	0.000	no trend
		Total organic carbon	01-01-1989	12-31-2009	36	4	0.509	0.200	0.100	3.30	0.697	0.003	0.020	+
		Bromide	01-01-1989	12-31-2009	24	2	0.023	0.020	0.010	0.040	0.008	0.252	0.000	no trend
USGS 18	434540112440901	Orthophosphate	01-01-1990	12-31-2009	19	8	0.012	0.010	0.009	0.020	0.004	0.174	0.000	no trend
USGS 19	433426112575701	Orthophosphate	01-01-1990	12-31-2009	26	19	0.010	0.009	0.008	0.023	0.003	0.899	-0.000	no trend
		Chromium	01-01-1990	12-31-2009	24	14	3.20	3.00	1.752	14.0	1.54	0.684	-0.067	no trend
USGS 23	434055112595901	Total organic carbon	01-01-1990	12-31-2009	15	1	1.21	0.912	0.400	5.17	1.19	0.005	0.079	+
		Orthophosphate	01-01-1990	12-31-2009	26	16	0.011	NA	0.010	0.020	0.003	0.089	0.000	no trend
		Chromium	01-01-1990	12-31-2009	24	13	3.36	3.20	1.335	14.0	2.16	0.747	-0.113	no trend
		Total organic carbon	01-01-1990	12-31-2009	15	6	1.04	0.300	0.100	7.96	2.05	0.959	0.014	no trend

Table 4. Statistical summaries and trend analyses of censored water-quality results for selected constituents in water from wells and surface-water sites at and near the Idaho National Laboratory, Idaho, 1980–2009.—Continued

[**Local name:** Local well identifier used in this study. **Site identifier:** Unique numerical identifiers used to access well data (http://waterdata.usgs.gov/nwis). **Constituent:** Aluminum, antimony, arsenic, barium, cobalt, chromium, copper, manganese, molybdenum, nickel, lead, thallium, and zinc in micrograms per liter. Other constituents in milligrams per liter. **Trend:** +, increasing; –, decreasing. NA, not applicable]

Table 4   51

Local name	Site identifier	Constituent	Start date	End date	Sample size	Number of censored values	Mean	Median	Minimum	Maximum	Standard deviation	p value	Slope	Trend
				Wells—Continued										
USGS 26	43521211239400 1	Aluminum	01-01-1990	12-31-2009	24	2	4.05	3.43	2.39	10.0	1.32	0.010	-0.208	–
		Antimony	01-01-1990	12-31-2009	22	16	0.109	0.100	0.096	1.00	0.023	0.886	-0.003	no trend
		Barium	01-01-1990	12-31-2009	25	1	36.5	36.3	34.4	40.5	1.23	0.132	-0.076	no trend
		Chromium	01-01-1990	12-31-2009	25	2	3.11	2.69	1.80	6.00	1.07	0.000	-0.144	–
		Cobalt	01-01-1990	12-31-2009	25	16	0.079	0.076	0.022	3.00	0.058	0.956	-0.001	no trend
		Copper	01-01-1990	12-31-2009	25	17	0.450	0.366	0.214	10.0	0.358	0.499	-0.052	no trend
		Manganese	01-01-1990	12-31-2009	25	13	0.549	0.405	0.304	2.00	0.404	0.372	-0.024	no trend
		Nickel	01-01-1990	12-31-2009	25	14	0.490	0.393	0.060	10.0	0.523	0.899	-0.000	no trend
		Thallium	01-01-1990	12-31-2009	22	21	0.439	NA	0.040	0.500	NA	0.799	-0.022	no trend
		Zinc	01-01-1990	12-31-2009	25	9	15.1	1.45	0.481	331	66.6	0.046	-0.176	–
		Orthophosphate	01-01-1990	12-31-2009	26	18	0.011	NA	0.010	0.021	0.003	0.015	0.001	+
		Total organic carbon	01-01-1990	12-31-2009	14	4	0.410	0.300	0.200	1.08	0.263	0.151	0.023	no trend
		Molybdenum	01-01-1990	12-31-2009	25	2	2.88	2.88	2.66	10.0	0.100	0.018	-0.011	–
USGS 27	43485111232180 1	Orthophosphate	01-01-1990	12-31-2009	26	19	0.011	NA	0.010	0.020	0.002	0.047	0.001	+
		Chromium	01-01-1990	12-31-2009	24	7	6.17	5.30	2.00	14.0	2.85	0.758	0.063	no trend
USGS 29	434407112285101	Orthophosphate	01-01-1990	12-31-2009	20	5	0.014	0.011	0.009	0.029	0.006	0.065	0.000	no trend
USGS 31	434625112342101	Orthophosphate	01-01-1990	12-31-2009	20	12	0.012	0.009	0.009	0.027	0.005	0.361	0.000	no trend
USGS 32	434444112322101	Orthophosphate	01-01-1990	12-31-2009	19	10	0.013	0.010	0.010	0.027	0.005	0.010	0.001	+
USGS 83	433023112561501	Chromium	08-01-1980	12-31-2009	23	2	13.0	13.0	2.01	21.4	4.30	0.311	-0.115	no trend
		Orthophosphate	08-01-1980	12-31-2009	28	16	0.012	0.010	0.010	0.021	0.004	0.341	0.000	no trend
USGS 86	43293511308000 1	Orthophosphate	01-01-1987	12-31-2009	24	3	0.017	0.017	0.010	0.028	0.006	0.002	0.001	+
		Chromium	01-01-1987	12-31-2009	26	2	12.6	12.3	1.00	19.0	2.20	0.520	-0.067	no trend
		Total organic carbon	01-01-1987	12-31-2009	15	2	0.549	0.300	0.100	1.70	0.470	0.960	-0.001	no trend

Table 4. Statistical summaries and trend analyses of censored water-quality results for selected constituents in water from wells and surface-water sites at and near the Idaho National Laboratory, Idaho, 1980–2009.—Continued

[**Local name:** Local well identifier used in this study. **Site identifier:** Unique numerical identifiers used to access well data (http://waterdata.usgs.gov/nwis). **Constituent:** Aluminum, antimony, arsenic, barium, cobalt, chromium, copper, manganese, molybdenum, nickel, lead, thallium, and zinc in micrograms per liter. Other constituents in milligrams per liter. **Trend:** +, increasing; −, decreasing. NA, not applicable]

Local name	Site identifier	Constituent	Start date	End date	Sample size	Number of censored values	Mean	Median	Minimum	Maximum	Standard deviation	p value	Slope	Trend
				Wells—Continued										
USGS 97	43380711255150l	Aluminum	01-01-1986	12-31-2009	24	1	3.52	4.00	1.602	10.0	1.63	0.160	-0.126	no trend
		Antimony	01-01-1986	12-31-2009	23	13	0.138	0.129	0.109	1.00	0.036	0.653	-0.004	no trend
		Cadmium	01-01-1986	12-31-2009	25	18	0.059	0.046	0.020	1.00	0.050	0.930	-0.001	no trend
		Chromium	01-01-1986	12-31-2009	36	2	6.86	6.89	4.000	50.0	1.62	1.00	0.000	no trend
		Cobalt	01-01-1986	12-31-2009	23	13	0.127	0.111	0.020	1.00	0.108	0.653	-0.013	no trend
		Copper	01-01-1986	12-31-2009	23	10	0.743	0.534	0.200	1.77	0.533	0.282	0.015	no trend
		Lead	01-01-1986	12-31-2009	26	14	1.24	0.843	0.046	14.0	2.83	0.981	0.002	no trend
		Manganese	01-01-1986	12-31-2009	24	18	0.158	0.115	0.069	1.00	0.211	0.923	-0.003	no trend
		Mercury	01-01-1986	12-31-2009	26	25	0.100	NA	0.010	0.230	NA	0.621	-0.007	no trend
		Nickel	01-01-1986	12-31-2009	23	10	0.855	0.406	0.060	3.09	0.894	0.617	0.040	no trend
		Selenium	01-01-1986	12-31-2009	25	1	2.01	2.00	0.129	2.60	0.292	0.615	-0.000	no trend
		Silver	01-01-1986	12-31-2009	25	24	1.00	NA	0.008	1.00	NA	0.462	-0.343	no trend
		Zinc	01-01-1986	12-31-2009	23	2	94.9	114	0.590	145	46.0	0.090	-2.87	no trend
		Fluoride	01-01-1986	12-31-2009	28	2	0.171	NA	0.100	0.200	0.046	0.563	0.000	no trend
		Orthophosphate	01-01-1986	12-31-2009	42	1	0.023	0.020	0.010	0.040	0.006	0.043	0.000	+
		Total organic carbon	01-01-1986	12-31-2009	43	1	0.765	0.500	0.200	9.17	1.35	0.006	0.019	+
USGS 98	43365711256360l	Aluminum	01-01-1986	12-31-2009	24	1	3.322	3.83	1.345	10.0	1.38	0.088	-0.166	no trend
		Antimony	01-01-1986	12-31-2009	23	15	0.113	0.103	0.092	1.00	0.031	0.923	0.001	no trend
		Arsenic	01-01-1986	12-31-2009	26	2	1.62	1.72	1.000	2.00	0.368	0.653	0.000	no trend
		Cadmium	01-01-1986	12-31-2009	26	20	0.120	0.043	0.020	1.00	0.143	0.636	-0.027	no trend
		Chromium	01-01-1986	12-31-2009	36	3	6.09	6.00	1.000	50.0	1.39	0.014	0.086	+
		Cobalt	01-01-1986	12-31-2009	23	14	0.081	0.071	0.020	1.00	0.078	0.700	-0.008	no trend
		Copper	01-01-1986	12-31-2009	23	5	1.93	1.32	0.250	9.70	2.32	0.769	0.000	no trend
		Lead	01-01-1986	12-31-2009	26	5	3.91	3.29	0.080	10.00	2.97	0.000	-0.525	−
		Manganese	01-01-1986	12-31-2009	24	13	1.37	0.512	0.343	10.22	2.27	0.132	0.194	no trend
		Mercury	01-01-1986	12-31-2009	24	23	0.100	NA	0.013	0.230	NA	0.929	-0.004	no trend
		Nickel	01-01-1986	12-31-2009	23	10	0.876	0.970	0.060	2.03	0.657	0.354	0.028	no trend
		Selenium	01-01-1986	12-31-2009	26	6	1.24	1.18	1.000	2.60	0.326	0.009	0.029	+
		Silver	01-01-1986	12-31-2009	26	24	0.083	0.006	0.006	2.00	0.542	0.542	-0.100	no trend
		Fluoride	01-01-1986	12-31-2009	25	2	0.208	0.200	0.100	0.400	0.070	0.258	-0.000	no trend
		Orthophosphate	01-01-1986	12-31-2009	38	4	0.015	0.016	0.009	0.024	0.005	0.990	-0.000	no trend
		Total organic carbon	01-01-1986	12-31-2009	41	1	3.13	0.400	0.100	78.2	12.8	0.000	0.070	+
USGS 99	43370511255210l	Orthophosphate	01-01-1986	12-31-2009	18	3	0.017	NA	0.010	0.020	0.005	0.559	0.000	no trend
		Chromium	01-01-1986	12-31-2009	36	8	5.85	5.69	3.000	50.0	1.97	0.521	0.020	no trend
		Fluoride	01-01-1986	12-31-2009	27	1	0.159	0.200	0.100	0.300	0.064	0.299	0.000	no trend
USGS 100	43350311240070l	Chromium	01-01-1986	12-31-2009	36	18	2.50	2.00	1.000	50.0	1.75	0.764	0.010	no trend

Table 4   53

Table 4. Statistical summaries and trend analyses of censored water-quality results for selected constituents in water from wells and surface-water sites at and near the Idaho National Laboratory, Idaho, 1980–2009.—Continued

[**Local name:** Local well identifier used in this study. **Site identifier:** Unique numerical identifiers used to access well data (http://waterdata.usgs.gov/nwis). **Constituent:** Aluminum, antimony, arsenic, barium, cobalt, chromium, copper, manganese, molybdenum, nickel, lead, thallium, and zinc in micrograms per liter. Other constituents in milligrams per liter. **Trend:** +, increasing; –, decreasing. NA, not applicable]

Local name	Site identifier	Constituent	Start date	End date	Sample size	Number of censored values	Mean	Median	Minimum	Maximum	Standard deviation	p value	Slope	Trend
				Wells—Continued										
USGS 101	43325511238801	Orthophosphate	01-01-1986	12-31-2009	24	13	0.011	0.010	0.009	0.020	0.003	0.289	0.000	no trend
		Chromium	01-01-1986	12-31-2009	33	23	1.39	0.877	0.698	50.0	1.42	0.921	-0.008	no trend
		Total organic carbon	01-01-1986	12-31-2009	14	2	0.851	0.526	0.190	3.84	0.957	0.824	-0.021	no trend
USGS 102	43385311255601	Orthophosphate	01-01-1990	12-31-2009	33	3	0.024	0.020	0.010	0.126	0.020	0.007	0.001	+
		Fluoride	01-01-1990	12-31-2009	25	1	0.192	NA	0.100	0.200	0.028	1.00	-0.000	no trend
USGS 103	43271411256070l	Orthophosphate	01-01-1986	12-31-2005	21	10	0.011	0.009	0.009	0.020	0.003	0.795	-0.000	no trend
		Chromium	01-01-1986	12-31-2005	22	8	6.70	6.400	5.00	14.0	1.44	0.099	0.200	no trend
		Total organic carbon	01-01-1986	12-31-2005	10	1	1.21	0.345	0.200	5.67	1.74	0.235	0.071	no trend
USGS 107	43294211253280l	Orthophosphate	01-01-1983	12-31-2009	30	13	0.011	0.010	0.009	0.020	0.004	0.131	0.000	no trend
		Chromium	01-01-1983	12-31-2009	24	8	6.21	5.215	4.02	14.0	2.50	0.487	-0.053	no trend
		Total organic carbon	01-01-1983	12-31-2009	14	1	0.517	0.395	0.178	1.20	0.344	0.739	0.005	no trend
USGS 109	43270111302560l	Orthophosphate	07-01-1987	12-31-2009	24	12	0.012	0.010	0.009	0.020	0.004	0.188	0.001	no trend
		Chromium	07-01-1987	12-31-2009	26	7	6.15	5.358	4.00	14.0	1.88	0.229	0.075	no trend
		Total organic carbon	07-01-1987	12-31-2009	15	1	0.682	0.745	0.211	1.19	0.341	0.426	0.023	no trend
USGS 110A	43271711250150l2	Orthophosphate	01-01-1995	12-31-2009	19	13	0.010	0.009	0.004	0.070	0.017	0.815	0.000	no trend
		Chromium	01-01-1995	12-31-2009	18	11	3.87	2.801	1.20	14.0	2.84	0.188	-0.454	no trend
		Total organic carbon	01-01-1995	12-31-2009	13	1	1.05	0.785	0.315	3.50	0.873	0.624	-0.042	no trend
USGS 117	43295511302590l	Orthophosphate	01-01-1987	12-31-2009	24	17	0.010	0.009	0.009	0.020	0.003	0.124	0.000	no trend
USGS 119	43294511302340l	Orthophosphate	01-01-1987	12-31-2009	24	12	0.010	0.009	0.008	0.020	0.002	0.690	0.000	no trend
USGS 121	43345011256030l	Orthophosphate	01-01-1991	12-31-2009	21	2	0.017	0.017	0.010	0.030	0.005	0.223	0.000	no trend
USGS 125	43260211305280l	Orthophosphate	01-01-1995	12-31-2009	23	7	0.011	0.010	0.009	0.020	0.003	0.524	0.000	no trend
		Chromium	01-01-1995	12-31-2009	23	9	4.94	5.000	2.30	14.0	2.53	0.615	-0.140	no trend
		Total organic carbon	01-01-1995	12-31-2009	15	1	0.972	0.861	0.200	3.18	0.827	0.090	0.089	no trend
USGS 126A	43552911247130l	Orthophosphate	01-01-2000	12-31-2009	11	7	0.011	0.011	0.010	0.020	0.002	0.623	0.000	no trend
		Chromium	01-01-2000	12-31-2009	12	5	2.48	2.114	1.20	10.0	1.51	0.753	-0.152	no trend
		Total organic carbon	01-01-2000	12-31-2009	9	3	1.04	0.615	0.150	4.51	1.40	0.668	0.098	no trend
USGS 126B	43552911247140l	Orthophosphate	01-01-2000	12-31-2009	11	7	0.012	0.011	0.010	0.020	0.002	0.922	0.000	no trend
		Chromium	01-01-2000	12-31-2009	11	4	2.06	1.910	1.14	10.0	0.879	0.482	-0.097	no trend
		Total organic carbon	01-01-2000	12-31-2009	9	3	0.596	0.396	0.300	1.46	0.430	0.357	0.051	no trend
WS INEL 1	43371611256360l	Orthophosphate	01-01-1986	12-31-2009	19	5	0.012	0.010	0.010	0.020	0.004	0.173	-0.000	no trend
		Chromium	01-01-1986	12-31-2009	34	3	9.04	8.400	5.49	50.0	2.80	0.020	-0.165	–
		Fluoride	01-01-1986	12-31-2009	26	3	0.154	0.100	0.100	0.500	0.091	0.372	-0.000	no trend
				Surface-water site										
Little Lost River near Howe	13119000	Fluoride	01-01-1965	12-31-2009	34	2	0.162	0.100	0.100	0.500	0.085	0.119	-0.000	no trend

Table 5. Statistical summaries and trend analyses for field measurements of pH, specific conductance, and water temperature for wells after installation of dedicated pumps and surface-water sites at and near the Idaho National Laboratory, Idaho, 1965–2009.

[All values are uncensored. **Local name:** Local well identifier used in this study. **Site identifier:** Unique numerical identifiers used to access well data (http://waterdata.usgs.gov/nwis). **Field measurement:** pH is in standard units. Specific conductance is in microsiemens per centimeter at 25 degrees Celsius. Temperature is in degrees Celsius. **Trend:** +, increasing; –, decreasing]

Local name	Site identifier	Field measurement	Start date	End date	Number of samples	Mean	Median	Minimum	Maximum	Standard deviation	p_value	Slope	Trend
				Wells									
ANP 6	43515211244 3101	pH	01-01-1986	12-31-2009	22	7.9	7.9	7.4	8.1	0.18	0.17	-0.01	no trend
		Specific conductance	01-01-1986	12-31-2009	24	400	404	300	426	26.2	0.119	1.723	no trend
		Temperature	01-01-1986	12-31-2009	23	13.5	13.4	12.9	15.0	0.439	0.279	0.017	no trend
ANP 9	43485611240 0001	pH	01-01-1994	12-31-2009	24	8.0	8.0	7.7	8.2	0.14	0.02	-0.02	–
		Specific conductance	01-01-1994	12-31-2009	24	378	379	350	398	9.28	0.624	0.164	no trend
		Temperature	01-01-1994	12-31-2009	24	14.2	14.2	13.9	14.4	0.128	0.008	0.018	+
ARBOR Test	43350911238 4801	pH	01-01-1989	12-31-2009	30	8.1	8.1	7.9	8.3	0.10	0.02	-0.01	–
		Specific conductance	01-01-1989	12-31-2009	30	330	332	310	348	9.96	0.000	1.067	+
		Temperature	01-01-1989	12-31-2009	30	13.6	13.5	13.2	14.5	0.283	0.528	0.000	no trend
AREA 2	43322311247 0201	pH	01-01-1992	12-31-2009	18	7.9	7.9	7.8	8.1	0.08	0.04	-0.01	–
		Specific conductance	01-01-1992	12-31-2009	18	362	364	351	366	4.20	0.349	-0.120	no trend
		Temperature	01-01-1992	12-31-2009	18	14.5	14.5	14.0	15.0	0.307	0.449	-0.012	no trend
Atomic City	43263811248 4101	pH	01-01-1980	12-31-2009	40	8.1	8.1	7.4	8.3	0.17	0.01	-0.01	–
		Specific conductance	01-01-1980	12-31-2009	42	350	354	290	366	15.7	0.000	0.932	+
		Temperature	01-01-1980	12-31-2009	41	13.7	13.9	9.10	21.1	2.72	0.135	-0.110	no trend
Badging Facility Well	43304211253 5101	pH	01-01-1985	12-31-2009	31	8.0	8.0	7.3	8.2	0.19	0.86	0.00	no trend
		Specific conductance	01-01-1985	12-31-2009	32	349	352	305	373	12.3	0.316	-0.347	no trend
		Temperature	01-01-1985	12-31-2009	34	12.5	12.5	11.0	14.6	0.716	0.190	0.027	no trend
CPP 4	43344011255 4401	pH	01-01-1983	12-31-2009	47	8.0	8.0	7.5	8.4	0.15	0.64	-0.00	no trend
		Specific conductance	01-01-1983	12-31-2009	50	405	410	320	460	29.2	0.633	0.589	no trend
		Temperature	01-01-1983	12-31-2009	53	12.6	12.5	11.0	17.0	1.02	0.032	-0.030	–
EBR 1	43305111300 2601	pH	01-01-1980	12-31-2009	38	8.2	8.2	7.7	8.4	0.13	0.32	-0.00	no trend
		Specific conductance	01-01-1980	12-31-2009	42	276	277	250	315	10.5	0.791	0.000	no trend
		Temperature	01-01-1980	12-31-2009	45	17.5	18.4	10.3	19.5	2.11	0.878	0.000	no trend
Fire Station 2	43354811256 2301	pH	01-01-1980	12-31-1996	47	8.0	8.0	7.4	8.3	0.15	0.97	0.00	no trend
		Specific conductance	01-01-1980	12-31-1996	52	428	429	360	456	18.9	0.396	-0.826	–
		Temperature	01-01-1980	12-31-1996	52	12.4	12.0	10.3	19.6	1.8	0.000	-0.134	–
Highway 3	43325611300 2501	pH	01-01-1980	12-31-2009	38	7.9	7.9	7.5	8.4	0.18	0.17	-0.01	no trend
		Specific conductance	01-01-1980	12-31-2009	40	342	345	280	420	19.6	0.558	-0.145	no trend
		Temperature	01-01-1980	12-31-2009	39	12.1	11.5	10.5	17.0	1.49	0.304	-0.026	no trend
IET 1 Disp	43515311242 0501	pH	01-01-1986	12-31-2009	20	7.8	7.8	7.5	7.9	0.13	0.00	-0.02	–
		Specific conductance	01-01-1986	12-31-2009	21	435	431	370	498	26.0	0.362	1.199	no trend
		Temperature	01-01-1986	12-31-2009	21	13.7	13.8	12.5	15.2	0.706	0.022	-0.055	–
Leo Rogers 1	43253311250 4901	pH	01-01-1980	12-31-2009	14	7.9	8.0	7.6	8.1	0.15	0.01	-0.02	–
		Specific conductance	01-01-1980	12-31-2009	19	372	372	360	380	4.91	0.357	-0.201	no trend
		Temperature	01-01-1980	12-31-2009	21	14.1	14.1	11.5	15.0	0.802	0.005	-0.055	–

Table 5   55

Table 5. Statistical summaries and trend analyses for field measurements of pH, specific conductance, and water temperature for wells after installation of dedicated pumps and surface-water sites at and near the Idaho National Laboratory, Idaho, 1965–2009—Continued.

[All values are uncensored. **Local name:** Local well identifier used in this study. **Site identifier:** Unique numerical identifiers used to access well data (http://waterdata.usgs.gov/nwis). **Field measurement:** pH is in standard units. Specific conductance is in microsiemens per centimeter at 25 degrees Celsius. Temperature is in degrees Celsius. **Trend:** +, increasing; –, decreasing]

Local name	Site identifier	Field measurement	Start date	End date	Number of samples	Mean	Median	Minimum	Maximum	Standard deviation	p_value	Slope	Trend
				Wells—Continued									
No Name 1	43503811245 3401	pH	01-01-1990	12-31-2009	27	8.0	8.1	7.5	8.2	0.16	0.03	-0.01	–
		Specific conductance	01-01-1990	12-31-2009	27	351	351	340	360	4.40	0.624	0.114	no trend
		Temperature	01-01-1990	12-31-2009	27	10.4	10.4	10.1	11.5	0.255	0.486	-0.007	no trend
NPR Test	43344911252 3101	pH	01-01-1986	12-31-2007	27	8.0	8.0	7.5	8.2	0.18	0.69	0.00	no trend
		Specific conductance	01-01-1986	12-31-2007	27	380	375	315	442	27.5	0.000	-4.057	–
		Temperature	01-01-1986	12-31-2007	28	12.1	12.0	10.6	14.5	0.672	0.896	0.000	no trend
P and W 2	43541911245 3101	pH	01-01-1986	12-31-2009	43	8.0	8.0	7.4	8.2	0.17	0.75	0.00	no trend
		Specific conductance	01-01-1986	12-31-2009	44	357	350	300	574	39.1	0.863	0.000	no trend
		Temperature	01-01-1986	12-31-2009	44	9.27	9.50	7.60	11.5	0.944	0.000	-0.116	–
PSTF Test	43494111245 4201	pH	01-01-1992	12-31-2009	26	8.1	8.1	7.9	8.2	0.11	0.02	-0.01	–
		Specific conductance	01-01-1992	12-31-2009	26	300	302	287	312	5.68	0.174	-0.398	no trend
		Temperature	01-01-1992	12-31-2009	26	13.5	13.4	13.0	13.6	0.158	0.639	0.000	no trend
Site 4	43361711254 2001	pH	01-01-1980	12-31-2009	25	8.0	8.0	7.5	8.2	0.16	0.90	0.00	no trend
		Specific conductance	01-01-1980	12-31-2009	25	402	387	336	535	56.4	0.000	-9.149	–
		Temperature	01-01-1980	12-31-2009	25	11.6	11.5	11.1	12.2	0.311	0.197	0.022	no trend
Site 9	43312311253 0101	pH	10-01-1990	12-31-2009	24	8.0	8.0	7.6	8.2	0.13	0.23	-0.01	no trend
		Specific conductance	10-01-1990	12-31-2009	24	352	353	334	367	8.37	0.017	0.805	+
		Temperature	10-01-1990	12-31-2009	24	14.0	14.0	13.3	14.5	0.289	0.269	-0.018	no trend
Site 14	43433411246 3101	pH	01-01-1980	12-31-2009	36	7.9	8.0	7.3	8.1	0.15	0.16	-0.00	no trend
		Specific conductance	01-01-1980	12-31-2009	36	330	332	270	341	12.0	0.604	0.059	no trend
		Temperature	01-01-1980	12-31-2009	38	16.7	16.6	16.0	17.5	0.288	0.564	0.000	no trend
Site 17	43402711257 5701	pH	01-01-1990	12-31-2009	20	7.8	7.8	7.3	7.9	0.13	0.01	-0.01	–
		Specific conductance	01-01-1990	12-31-2009	20	419	418	374	458	20.0	0.155	-1.341	no trend
		Temperature	01-01-1990	12-31-2009	20	12.5	12.5	12.0	13.3	0.335	0.015	0.035	+
Site 19	43352211258 2101	pH	01-01-1986	12-31-2009	32	7.9	7.9	7.4	8.1	0.19	0.01	-0.01	–
		Specific conductance	01-01-1986	12-31-2009	34	391	387	340	440	20.0	0.027	-0.749	–
		Temperature	01-01-1986	12-31-2009	34	14.5	14.8	13.0	15.3	0.728	0.224	0.025	no trend
SPERT 1	43325211252 0301	pH	01-01-1980	12-31-2009	31	8.0	8.0	7.6	8.2	0.14	0.17	-0.01	no trend
		Specific conductance	01-01-1980	12-31-2009	32	399	390	325	485	36.3	0.082	1.402	–
		Temperature	01-01-1980	12-31-2009	36	12.1	12.0	11.2	16.7	0.966	0.035	-0.029	–
TRA 1	43352111257 3801	pH	01-01-1980	12-31-2009	18	7.8	7.8	7.6	8.1	0.15	0.22	-0.01	no trend
		Specific conductance	01-01-1980	12-31-2009	30	406	412	325	428	21.5	0.003	-1.006	–
		Temperature	01-01-1980	12-31-2009	29	13.9	13.6	13.0	17.3	1.025	0.262	-0.018	no trend
TRA 3	43352211257 3501	pH	01-01-1980	12-31-2009	25	7.9	7.9	7.4	8.2	0.18	0.10	-0.01	no trend
		Specific conductance	01-01-1980	12-31-2009	30	419	420	350	442	18.2	0.003	-1.246	–
		Temperature	01-01-1980	12-31-2009	29	13.6	13.2	12.8	17.6	1.07	0.083	-0.031	no trend

Table 5. Statistical summaries and trend analyses for field measurements of pH, specific conductance, and water temperature for wells after installation of dedicated pumps and surface-water sites at and near the Idaho National Laboratory, Idaho, 1965–2009—Continued.

[All values are uncensored. **Local name:** Local well identifier used in this study. **Site identifier:** Unique numerical identifiers used to access well data (http://waterdata.usgs.gov/nwis). **Field measurement:** pH is in standard units. Specific conductance is in microsiemens per centimeter at 25 degrees Celsius. Temperature is in degrees Celsius. **Trend:** +, increasing; –, decreasing]

Local name	Site identifier	Field measurement	Start date	End date	Number of samples	Mean	Median	Minimum	Maximum	Standard deviation	p_value	Slope	Trend
				Wells—Continued									
TRA 4	43352111257420 1	pH	01-01-1980	12-31-2009	28	7.8	7.9	7.5	8.3	0.17	0.02	-0.01	--
		Specific conductance	01-01-1980	12-31-2009	31	399	402	340	425	14.7	0.045	-0.803	--
		Temperature	01-01-1980	12-31-2009	30	14.59	14.4	13.8	18.8	1.10	0.234	-0.020	no trend
USGS 1	43270011247080 1	pH	01-01-1990	12-31-2009	24	8.0	8.0	7.5	8.1	0.18	0.00	-0.02	--
		Specific conductance	01-01-1990	12-31-2009	26	318	320	303	333	6.21	0.000	0.670	+
		Temperature	01-01-1990	12-31-2009	26	14.3	14.4	12.3	15.0	0.46	0.873	0.000	no trend
USGS 2	43332011243230 1	pH	01-01-1990	12-31-2009	19	7.9	8.0	7.7	8.1	0.13	0.19	-0.01	no trend
		Specific conductance	01-01-1990	12-31-2009	19	351	354	334	362	7.62	0.175	0.339	no trend
		Temperature	01-01-1990	12-31-2009	19	13.7	13.7	13.0	14.3	0.306	0.659	-0.012	no trend
USGS 4	43465711228220 1	pH	01-01-1990	12-31-2009	26	7.6	7.6	7.1	7.8	0.15	0.00	-0.02	--
		Specific conductance	01-01-1990	12-31-2009	26	715	721	685	744	17.1	0.185	-1.605	no trend
		Temperature	01-01-1990	12-31-2009	26	11.5	11.4	11.0	12.5	0.427	0.646	0.000	no trend
USGS 5	43354311249380 1	pH	01-01-1990	12-31-2009	26	7.9	7.9	7.4	8.1	0.21	0.13	-0.02	--
		Specific conductance	01-01-1990	12-31-2009	26	322	331	262	338	21.9	0.002	-1.135	--
		Temperature	01-01-1990	12-31-2009	26	14.6	14.8	12.6	15.5	0.651	0.250	-0.043	no trend
USGS 6	43403111245370 1	pH	01-01-1990	12-31-2009	19	7.9	7.9	7.4	8.2	0.25	0.00	-0.03	--
		Specific conductance	01-01-1990	12-31-2009	19	301	295	280	357	16.6	0.015	1.173	+
		Temperature	01-01-1990	12-31-2009	19	14.2	14.1	13.5	15.0	0.351	0.132	-0.024	no trend
USGS 7	43491511244390 1	pH	01-01-1990	12-31-2009	26	8.0	8.1	7.7	8.2	0.13	0.74	-0.00	no trend
		Specific conductance	01-01-1990	12-31-2009	26	303	305	290	308	4.65	0.863	0.000	no trend
		Temperature	01-01-1990	12-31-2009	26	19.4	19.4	18.8	20.0	0.242	0.788	0.000	no trend
USGS 8	43312111311580 1	pH	01-01-1990	12-31-2009	32	8.0	8.0	7.6	8.4	0.15	0.02	-0.01	--
		Specific conductance	01-01-1990	12-31-2009	32	366	370	295	378	15.2	0.870	0.000	no trend
		Temperature	01-01-1990	12-31-2009	32	11.2	11.1	10.6	12.0	0.343	0.917	0.000	no trend
USGS 9	43274011304450 1	pH	01-01-1987	12-31-2009	42	8.1	8.2	7.7	9.0	0.22	0.00	-0.02	--
		Specific conductance	01-01-1987	12-31-2009	42	384	383	233	465	32.8	0.000	-1.995	--
		Temperature	01-01-1987	12-31-2009	42	11.4	11.4	10.0	12.5	0.407	0.150	-0.017	no trend
USGS 11	43233611306420 1	pH	09-01-1989	12-31-2009	42	8.0	8.1	7.7	8.2	0.11	0.24	-0.00	no trend
		Specific conductance	09-01-1989	12-31-2009	42	354	354	332	370	7.22	0.002	-0.620	--
		Temperature	09-01-1989	12-31-2009	42	12.0	11.9	10.5	13.0	0.442	0.120	-0.017	--
USGS 12	43412611255070 1	pH	01-01-1990	12-31-2009	81	7.8	7.8	7.4	8.0	0.14	0.01	-0.01	--
		Specific conductance	01-01-1990	12-31-2009	81	540	540	461	610	45.7	0.000	-4.335	--
		Temperature	01-01-1990	12-31-2009	81	11.8	11.7	11.0	14.0	0.478	0.200	-0.018	no trend
USGS 14	43201911256320 1	pH	01-01-1989	12-31-2009	42	8.0	8.0	7.5	8.2	0.15	0.01	-0.01	--
		Specific conductance	01-01-1989	12-31-2009	42	382	385	268	400	19.9	0.000	0.868	+
		Temperature	01-01-1989	12-31-2009	42	14.9	14.8	14.0	16.2	0.434	0.063	-0.026	no trend

Table 5   57

Table 5. Statistical summaries and trend analyses for field measurements of pH, specific conductance, and water temperature for wells after installation of dedicated pumps and surface-water sites at and near the Idaho National Laboratory, Idaho, 1965–2009—Continued.

[All values are uncensored. **Local name:** Local well identifier used in this study. **Site identifier:** Unique numerical identifiers used to access well data (http://waterdata.usgs.gov/nwis). **Field measurement**: pH is in standard units. Specific conductance is in microsiemens per centimeter at 25 degrees Celsius. Temperature is in degrees Celsius. **Trend:** +, increasing; –, decreasing]

Local name	Site identifier	Field measurement	Start date	End date	Number of samples	Mean	Median	Minimum	Maximum	Standard deviation	p_value	Slope	Trend
		Wells—Continued											
USGS 15	43423411255701	pH	01-01-1990	12-31-2009	39	7.9	7.9	7.2	8.2	0.18	0.00	-0.02	–
		Specific conductance	01-01-1990	12-31-2009	40	416	396	298	658	96.6	0.080	3.87	no trend
		Temperature	01-01-1990	12-31-2009	40	11.3	11.4	10.2	12.0	0.460	0.000	-0.045	no trend
USGS 17	43393711251540 1	pH	01-01-1989	12-31-2009	45	8.1	8.2	7.3	8.3	0.16	0.37	-0.00	no trend
		Specific conductance	01-01-1989	12-31-2009	45	297	298	278	307	6.11	0.058	0.413	no trend
		Temperature	01-01-1989	12-31-2009	45	13.2	13.0	12.0	14.1	0.431	0.598	0.000	no trend
USGS 18	43454011244090 1	pH	01-01-1990	12-31-2009	19	7.9	8.0	7.5	8.1	0.20	0.00	-0.02	+
		Specific conductance	01-01-1990	12-31-2009	19	344	345	313	350	8.34	0.003	0.614	no trend
		Temperature	01-01-1990	12-31-2009	19	15.4	15 5	14.5	16.0	0.315	0.147	0.025	–
USGS 19	43442611257570 1	pH	01-01-1990	12-31-2009	32	7.7	7.7	7.2	7.9	0.18	0.02	-0.01	–
		Specific conductance	01-01-1990	12-31-2009	32	397	396	368	441	15.3	0.000	-1 50	no trend
		Temperature	01-01-1990	12-31-2009	32	17.2	17 2	16.5	17.8	0.283	0.080	0.025	no trend
USGS 22	43342211303170 1	pH	01-01-1990	12-31-2009	23	8.0	8.0	7.7	8.2	0.13	0.17	-0.01	no trend
		Specific conductance	01-01-1990	12-31-2009	23	402	406	359	419	15.7	0.124	0.717	no trend
		Temperature	01-01-1990	12-31-2009	23	18.3	18.8	14.5	20.1	1.38	0.856	0.000	no trend
USGS 23	43405511259590 1	pH	01-01-1990	12-31-2009	26	7.9	7 9	7.4	8.1	0.14	0.38	-0.01	no trend
		Specific conductance	01-01-1990	12-31-2009	26	356	354	333	380	12.0	0.379	-0.537	no trend
		Temperature	01-01-1990	12-31-2009	26	15.7	15.6	15.1	16.4	0.368	0.220	0.029	no trend
USGS 26	43521212394001	pH	01-01-1990	12-31-2009	27	7.8	7 9	7.3	8.0	0.16	0.05	-0.01	no trend
		Specific conductance	01-01-1990	12-31-2009	27	387	390	365	396	8.10	0.197	0.412	no trend
		Temperature	01-01-1990	12-31-2009	27	15.0	15.0	14.5	15.5	0.253	0.125	0.020	no trend
USGS 27	43485111232180 1	pH	01-01-1990	12-31-2009	32	7.9	8.0	7.5	8.1	0.13	0.00	-0.01	–
		Specific conductance	01-01-1990	12-31-2009	32	556	561	526	570	11.0	0.326	0.489	no trend
		Temperature	01-01-1990	12-31-2009	32	15.7	15.8	14.5	16.5	0.408	0.361	-0.015	no trend
USGS 29	43440711228510 1	pH	01-01-1990	12-31-2009	20	7.8	7 9	7.3	8.0	0.24	0.00	-0.02	–
		Specific conductance	01-01-1990	12-31-2009	20	436	434	411	470	19.4	0.053	-1.74	no trend
		Temperature	01-01-1990	12-31-2009	20	12.7	12.7	12.1	13.5	0.296	0.953	0.000	no trend
USGS 31	43462511234210 1	pH	01-01-1990	12-31-2009	20	7.9	7 9	7.6	8.1	0.12	0.07	-0.01	no trend
		Specific conductance	01-01-1990	12-31-2009	20	413	414	387	432	14.7	0.000	2.66	+
		Temperature	01-01-1990	12-31-2009	20	15.6	15 5	15.0	16.0	0.274	0.394	-0.013	no trend
USGS 32	43444411232210 1	pH	01-01-1990	12-31-2009	20	7.8	7 9	7.4	8.0	0.14	0.01	-0.01	–
		Specific conductance	01-01-1990	12-31-2009	20	495	484	429	577	51.4	0.162	-2.13	no trend
		Temperature	01-01-1990	12-31-2009	20	14.6	14 5	14.0	15.5	0.372	0.830	0.000	no trend
USGS 83	43302311256150 1	pH	08-01-1980	12-31-2009	38	8.1	8 1	7.4	8.3	0.21	0.83	0.00	no trend
		Specific conductance	08-01-1980	12-31-2009	40	273	275	220	291	12.6	0.818	0.000	no trend
		Temperature	08-01-1980	12-31-2009	47	12.0	11 9	11.2	13.5	0.44	0.002	-0.020	–

Table 5. Statistical summaries and trend analyses for field measurements of pH, specific conductance, and water temperature for wells after installation of dedicated pumps and surface-water sites at and near the Idaho National Laboratory, Idaho, 1965–2009—Continued.

[All values are uncensored. **Local name:** Local well identifier used in this study. **Site identifier:** Unique numerical identifiers used to access well data (http://waterdata.usgs.gov/nwis). **Field measurement:** pH is in standard units. Specific conductance is in microsiemens per centimeter at 25 degrees Celsius. Temperature is in degrees Celsius. **Trend** +, increasing; –, decreasing]

Local name	Site identifier	Field measurement	Start date	End date	Number of samples	Mean	Median	Minimum	Maximum	Standard deviation	p_value	Slope	Trend
				Wells—Continued									
USGS 86	43293511308001	pH	01-01-1987	12-31-2009	40	8.1	8.1	7.6	8.3	0.14	0.52	0.00	no trend
		Specific conductance	01-01-1987	12-31-2009	40	334	334	≤12	350	8.21	0.000	-0.831	–
		Temperature	01-01-1987	12-31-2009	40	10.2	10.0	9.00	14.0	0.753	0.566	0.000	no trend
USGS 97	433807112551501	pH	01-01-1986	12-31-2009	99	7.9	7.9	7.4	8.1	0.14	0.20	-0.00	no trend
		Specific conductance	01-01-1986	12-31-2009	101	581	580	470	615	23.7	0.828	0.083	no trend
		Temperature	01-01-1986	12-31-2009	102	11.6	11.5	10.1	12.6	0.360	0.038	-0.012	–
USGS 98	433657112563601	pH	01-01-1986	12-31-2009	97	7.9	8.0	7.5	8.1	0.14	0.00	-0.01	–
		Specific conductance	01-01-1986	12-31-2009	99	420	430	375	484	21.2	0.594	0.243	no trend
		Temperature	01-01-1986	12-31-2009	100	12.2	12.1	10.5	14.0	0.380	0.725	0.000	no trend
USGS 99	433705112552101	pH	01-01-1986	12-31-2009	101	7.9	7.9	7.5	8.1	0.14	0.02	-0.01	+
		Specific conductance	01-01-1986	12-31-2009	103	525	528	405	560	17.7	0.000	1.166	+
		Temperature	01-01-1986	12-31-2009	104	11.5	11.5	10.0	13.2	0.417	0.000	-0.025	–
USGS 100	433503112400701	pH	01-01-1986	12-31-2009	46	8.0	8.1	7.4	8.3	0.16	0.20	-0.01	no trend
		Specific conductance	01-01-1986	12-31-2009	49	349	352	280	383	17.3	0.000	1.32	+
		Temperature	01-01-1986	12-31-2009	50	13.7	13.5	13.1	15.0	0.336	0.626	0.000	no trend
USGS 101	433255112381801	pH	01-01-1986	12-31-2009	38	8.1	8.1	7.8	8.3	0.14	0.02	-0.01	–
		Specific conductance	01-01-1986	12-31-2009	40	278	281	240	295	10.3	0.000	0.796	+
		Temperature	01-01-1986	12-31-2009	41	13.9	13.9	12.5	14.9	0.402	0.624	0.000	no trend
USGS 102	433853112551601	pH	01-01-1990	12-31-2009	75	7.9	7.9	7.5	8.1	0.14	0.00	-0.01	–
		Specific conductance	01-01-1990	12-31-2009	75	571	565	538	606	20.3	0.097	-0.862	no trend
		Temperature	01-01-1990	12-31-2009	75	11.6	11.5	10.5	13.0	0.338	0.110	-0.014	no trend
USGS 103	432714112560701	pH	01-01-1986	12-31-2005	59	8.1	8.1	7.8	8.3	0.11	0.41	0.00	no trend
		Specific conductance	01-01-1986	12-31-2005	59	363	364	340	387	8.13	0.002	0.685	+
		Temperature	01-01-1986	12-31-2005	63	13.6	13.6	12.0	15.0	0.439	0.523	0.000	no trend
USGS 107	432942112532801	pH	01-01-1983	12-31-2009	40	8.0	8.0	7.7	8.4	0.13	0.00	-0.01	–
		Specific conductance	01-01-1983	12-31-2009	42	394	398	330	430	17.2	0.000	1.55	+
		Temperature	01-01-1983	12-31-2009	51	14.8	14.7	12.6	21.1	1.04	0.299	0.010	no trend
USGS 109	432701113025601	pH	07-01-1987	12-31-2009	40	8.0	8.1	7.5	8.2	0.16	0.02	-0.01	–
		Specific conductance	07-01-1987	12-31-2009	40	380	377	360	410	12.3	0.012	-0.821	–
		Temperature	07-01-1987	12-31-2009	40	13.6	13.6	12.2	14.5	0.430	0.017	-0.028	–
USGS 110	432717112501501	pH	01-01-1983	12-31-1992	20	8.1	8.1	7.9	8.3	0.10	0.93	0.00	no trend
		Specific conductance	01-01-1983	12-31-1992	22	372	370	315	419	20.3	0.174	1.99	no trend
		Temperature	01-01-1983	12-31-1992	32	15.0	14.5	13.0	29.8	3.16	0.830	0.000	no trend
USGS 110A	432717112501502	pH	01-01-1995	12-31-2009	19	7.9	7.9	7.4	8.1	0.18	0.00	-0.03	–
		Specific conductance	01-01-1995	12-31-2009	20	376	378	364	382	5.74	0.229	-0.345	no trend
		Temperature	01-01-1995	12-31-2009	20	14.4	14.3	14.0	15.8	0.466	0.057	0.051	no trend

Table 5. Statistical summaries and trend analyses for field measurements of pH, specific conductance, and water temperature for wells after installation of dedicated pumps and surface-water sites at and near the Idaho National Laboratory, Idaho, 1965–2009—Continued.

[All values are uncensored. **Local name:** Local well identifier used in this study. **Site identifier:** Unique numerical identifiers used to access well data (http://waterdata.usgs.gov/nwis). **Field measurement:** pH is in standard units. Specific conductance is in microsiemens per centimeter at 25 degrees Celsius. Temperature is in degrees Celsius. **Trend:** +, increasing; –, decreasing]

Local name	Site identifier	Field measurement	Start date	End date	Number of samples	Mean	Median	Minimum	Maximum	Standard deviation	p_value	Slope	Trend
				Wells—Continued									
USGS 117	432955113025901	pH	01-01-1987	12-31-2009	65	8.3	8.3	7.9	8.6	0.12	0.26	-0.00	no trend
		Specific conductance	01-01-1987	12-31-2009	65	271	275	234	282	10.1	0.845	0.000	no trend
		Temperature	01-01-1987	12-31-2009	65	13.5	13.4	12.7	15.7	0.551	0.003	-0.038	–
USGS 119	432945113023401	pH	01-01-1987	12-31-2009	66	8.4	8.4	7.9	9.1	0.22	0.52	-0.00	no trend
		Specific conductance	01-01-1987	12-31-2009	66	280	281	243	309	11.4	0.219	-0.398	no trend
		Temperature	01-01-1987	12-31-2009	66	14.4	14.5	12.1	16.2	0.801	0.135	-0.036	no trend
USGS 121	433450112560301	pH	01-01-1991	12-31-2009	32	7.9	8.0	7.5	8.4	0.18	0.01	-0.02	–
		Specific conductance	01-01-1991	12-31-2009	32	382	383	320	402	14.7	0.501	-0.393	no trend
		Temperature	01-01-1991	12-31-2009	32	11.7	11.6	11.0	13.0	0.365	0.382	0.013	no trend
USGS 125	432602113052801	pH	01-01-1995	12-31-2009	25	8.0	8.0	7.4	8.2	0.18	0.01	-0.02	–
		Specific conductance	01-01-1995	12-31-2009	25	370	370	354	386	8.59	0.000	-1.57	no trend
		Temperature	01-01-1995	12-31-2009	25	12.5	12.5	11.9	12.8	0.190	0.995	0.000	no trend
USGS 126A	435529112471301	pH	01-01-2000	12-31-2009	12	7.9	8.0	7.5	8.1	0.23	0.05	-0.03	no trend
		Specific conductance	01-01-2000	12-31-2009	12	347	346	335	353	5.23	0.756	-0.091	no trend
		Temperature	01-01-2000	12-31-2009	12	11.1	11.0	10.8	11.5	0.198	0.219	-0.025	no trend
USGS 126B	435529112471401	pH	01-01-2000	12-31-2009	12	8.3	8.3	8.2	8.4	0.07	0.90	0.00	no trend
		Specific conductance	01-01-2000	12-31-2009	12	344	344	337	350	3.79	0.449	-0.559	no trend
		Temperature	01-01-2000	12-31-2009	12	10.7	10.6	10.4	11.2	0.204	0.820	0.000	no trend
WS INEL 1	433716112563601	pH	01-01-1986	12-31-2009	55	7.9	7.9	7.4	8.2	0.15	0.78	-0.00	no trend
		Specific conductance	01-01-1986	12-31-2009	57	679	686	494	910	112	0.000	-19.4	–
		Temperature	01-01-1986	12-31-2009	57	12.1	12.0	10.3	13.6	0.529	0.955	0.000	no trend

Table 5   59

Table 5. Statistical summaries and trend analyses for field measurements of pH, specific conductance, and water temperature for wells after installation of dedicated pumps and surface-water sites at and near the Idaho National Laboratory, Idaho, 1965–2009—Continued.

[All values are uncensored. **Local name:** Local well identifier used in this study. **Site identifier:** Unique numerical identifiers used to access well data (http://waterdata.usgs.gov/nwis). **Field measurement:** pH is in standard units. Specific conductance is in microsiemens per centimeter at 25 degrees Celsius. Temperature is in degrees Celsius. **Trend:** +, increasing; –, decreasing]

Local name	Site identifier	Field measurement	Start date	End date	Number of samples	Mean	Median	Minimum	Maximum	Standard deviation	p_value	Slope	Trend
				Surface-water sites									
Big Lost River at Experimental Dairy Farm, near Howe	13132545	pH	01-01-1981	12-31-2009	13	8.3	8.5	7.3	8.7		0.44	-0.03	no trend
		Specific conductance	01-01-1981	12-31-2009	15	318	346	230	393		0.868	0.000	no trend
		Temperature	01-01-1981	12-31-2009	22	11.4	10.9	3.70	22.0		0.127	0.200	no trend
Big Lost River below INL diversion, near Arco	13132520	pH	01-01-1981	12-31-2009	19	8.3	8.4	7.6	8.7		0.78	-0.00	no trend
		Specific conductance	01-01-1981	12-31-2009	68	351	365	229	486		0.070	-1.35	no trend
		Temperature	01-01-1981	12-31-2009	78	9.82	9.00	-.00	26.0		0.038	0.225	+
Big Lost River below Mackay Reservoir, near Mackay	13127700	pH	01-01-1985	12-31-2009	34	8.6	8.6	7.5	9.0		0.20	0.01	no trend
		Specific conductance	01-01-1985	12-31-2009	34	295	294	275	321		0.000	-1.05	–
		Temperature	01-01-1985	12-31-2009	34	8.14	8.05	3.30	13.7		0.124	0.165	no trend
Big Lost River near Arco	13132500	pH	01-01-1965	12-31-2009	74	8.1	8.1	7.1	9.4		0.00	0.01	+
		Specific conductance	01-01-1965	12-31-2009	192	393	402	222	563		0.000	-1.89	–
		Temperature	01-01-1965	12-31-2009	191	9.54	9.10	-1.00	26.0		0.678	0.003	no trend
Birch Creek at Blue Dome, near Reno	13117020	pH	01-01-1970	12-31-2009	31	8.4	8.5	7.6	8.8		0.03	-0.02	–
		Specific conductance	01-01-1970	12-31-2009	123	330	332	111	586		0.235	0.259	no trend
		Temperature	01-01-1970	12-31-2009	146	8.09	8.00	0.000	19.0		0.000	-0.115	–
Little Lost River near Howe	13119000	pH	01-01-1965	12-31-2009	70	8.3	8.4	6.8	8.8		0.00	0.01	+
		Specific conductance	01-01-1965	12-31-2009	161	344	346	173	501		0.244	0.663	no trend
		Temperature	01-01-1965	12-31-2009	189	8.60	8.50	0.000	19.0		0.275	0.044	no trend
Mud Lake near Terreton	13115000	pH	01-01-1965	12-31-2009	58	8.4	8.4	7.6	9.5		0.01	0.02	+
		Specific conductance	01-01-1965	12-31-2009	69	276	288	162	334		0.002	1.17	+
		Temperature	01-01-1965	12-31-2009	93	10.5	10.4	0.000	26.0		0.334	0.035	no trend

Table 6   61

Table 6. Summary of water-quality results for selected radiochemical constituents in water in water from wells and surface-water sites at and near the Idaho National Laboratory, Idaho.

[Data are available at http://waterdata.usgs.gov/nwis. **Local name:** Local well identifier used in this study. BLR, Big Lost River; BC, Birch Creek; LLR, Little Lost River. **Constituent:** Concentrations are in picocuries per liter. **Remarks**: Concentrations are given for values greater than the reporting level. Uncertainty is 1 sample standard deviation]

Local name	Constituent	Number of samples	Number greater than reporting level	Period of record	Remarks
			Wells		
ANP 6	Tritium	22	0	1977–2009	
	Strontium-90	19	0	1987–2009	
ANP 9	Tritium	24	0	1977–2009	
	Strontium-90	23	0	1994–2009	
	Cesium-137	23	0	1994–2009	
	Gross alpha	23	0	1994–2009	
	Gross beta	23	0	1994–2009	
ARBOR Test	Tritium	33	0	1960–2009	
AREA 2	Tritium	22	0	1966–2009	
	Strontium-90	20	0	1990–2009	
Atomic City	Tritium	76	0	1966–2009	
Badging Facility Well	Tritium	33	0	1985–2009	
	Strontium-90	18	0	1985–2009	
CPP 4	Tritium	54	1	1983–2009	2,300 ±400 in 1983
	Strontium-90	55	0	1983–2009	
EBR1	Tritium	77	0	1966–2009	
	Cesium-137	25	0	1987–2009	
	Gross alpha	27	0	1977–2009	
	Gross beta	27	0	1977–2009	
Fire Station 2	Tritium	76	0	1972–1996	
Highway 3	Tritium	60	0	1975–2009	
	Cesium-137	22	0	1994–2009	
	Gross alpha	22	0	1994–2009	
	Gross beta	22	0	1994–2009	
IET 1 Disp	Tritium	24	0	1982–2009	
	Strontium-90	21	0	1987–2009	
Leo Rogers	Tritium	28	1	1971–2009	7,000 ±2,000 in 1971
	Gross Alpha	22	0	1971–2009	
	Gross Beta	22	0	1971–2009	
	Cesium-137	23	0	1980–2009	
No Name 1	Tritium	27	0	1991–2009	
	Strontium-90	27	0	1991–2009	
	Cesium-137	25	0	1991–2009	
	Gross alpha	24	0	1994–2009	
	Gross beta	24	0	1994–2009	
NPR Test	Tritium	31	0	1986–2009	
	Cesium-137	26	0	1986–2009	
	Gross alpha	23	0	1994–2009	
	Gross beta	23	0	1994–2009	
P and W2	Tritium	69	0	1960–2009	
	Cesium-137	26	0	1987–2009	
	Gross alpha	24	0	1994–2009	
	Gross beta	24	0	1994–2009	
PSTF Test	Tritium	28	0	1966–2009	
	Strontium-90	26	0	1990–2009	
	Cesium-137	23	0	1990–2009	
	Gross alpha	23	0	1994–2009	
	Gross beta	23	0	1994–2009	

Table 6. Summary of water-quality results for selected radiochemical constituents in water in water from wells and surface-water sites at and near the Idaho National Laboratory, Idaho.—Continued

[Data are available at http://waterdata.usgs.gov/nwis. **Local name:** Local well identifier used in this study. BLR, Big Lost River; BC, Birch Creek; LLR, Little Lost River. **Constituent:** Concentrations are in picocuries per liter. **Remarks:** Concentrations are given for values greater than the reporting level. Uncertainty is sample standard deviation]

Local name	Constituent	Number of samples	Number greater than reporting level	Period of record	Remarks
		Wells—Continued			
Site 4	Tritium	26	0	1977–2009	
Site 9	Tritium	76	0	1967–2009	
	Strontium-90	25	0	1990–2009	
Site 14	Tritium	44	0	1975–2009	
	Cesium-137	23	0	1991–2009	
	Gross alpha	22	0	1995–2009	
	Gross beta	22	0	1995–2009	
Site 17	Tritium	23	0	1977–2009	
	Strontium-90	19	0	1991–2009	
Site 19	Tritium	73	2	1972–2009	600 ±200 in 1975; 26,400 ±600 in 1979
SPERT 1	Tritium	49	0	1972–2009	
TRA 1	Tritium	64	0	1960–2009	
TRA 3	Tritium	62	0	1972–2009	
TRA 4	Tritium	64	0	1972–2009	
USGS 1	Tritium	30	1	1966–2009	3,000 ±1,000 in 1966
	Cesium-137	23	0	1991–2009	
	Gross alpha	23	0	1994–2009	
	Gross beta	24	0	1994–2009	
USGS 2	Tritium	20	0	1977–2009	
	Strontium-90	18	0	1991–2009	
USGS 4	Tritium	28	0	1977–2009	
	Cesium-137	24	0	1991–2009	
	Gross alpha	23	0	1994–2009	
	Gross beta	23	0	1994–2009	
USGS 5	Tritium	27	0	1977–2009	
	Cesium-137	24	0	1990–2009	
	Gross alpha	23	0	1994–2009	
	Gross beta	23	0	1994–2009	
USGS 6	Tritium	20	0	1977–2009	
	Strontium-90	19	0	1990–2009	
USGS 7	Tritium	29	0	1961–2009	
	Strontium-90	26	0	1991–2009	
	Cesium-137	24	0	1991–2009	
	Gross alpha	23	0	1994–2009	
	Gross beta	23	0	1994–2009	
USGS 8	Tritium	54	0	1966–2009	
	Cesium-137	47	0	1982–2009	
	Gross alpha	47	0	1971–2009	
	Gross beta	46	0	1980–2009	
USGS 9	Tritium	84	0	1966–2009	
	Cesium-137	28	0	1980–2009	
	Gross alpha	28	0	1969–2009	
	Gross beta	28	2	1969–2009	6 ±2 in 1994; 7 ±2 in 2003
USGS 11	Tritium	70	0	1960–2009	
	Cesium-137	53	0	1978–2009	
	Gross alpha	46	2	1971–2009	2.9 ±0.5 in 1980; 4 ±1 in 1982
	Gross beta	46	0	1971–2009	

Table 6   63

Table 6. Summary of water-quality results for selected radiochemical constituents in water in water from wells and surface-water sites at and near the Idaho National Laboratory, Idaho.—Continued

[Data are available at http://waterdata.usgs.gov/nwis. **Local name:** Local well identifier used in this study. BLR, Big Lost River; BC, Birch Creek; LLR, Little Lost River. **Constituent:** Concentrations are in picocuries per liter. **Remarks:** Concentrations are given for values greater than the reporting level. Uncertainty is sample standard deviation]

Local name	Constituent	Number of samples	Number greater than reporting level	Period of record	Remarks
		Wells—Continued			
USGS 12	Tritium	52	0	1977–2009	
	Cesium-137	27	0	1994–2009	
	Gross alpha	22	0	1994–2009	
	Gross beta	22	0	1994–2009	
USGS 14	Tritium	68	0	1966–2009	
	Cesium-137	54	0	1978–2009	
	Gross alpha	45	2	1971–2009	4,000 ±1000 in 1971; 6 ±2 in 1982
	Gross beta	45	0	1971–2009	
USGS 15	Tritium	30	0	1977–2009	
	Strontium-90	19	0	1991–2009	
USGS 17	Tritium	35	0	1977–2009	
	Cesium-137	26	0	1990–2009	
	Gross alpha	23	0	1994–2009	
	Gross beta	23	0	1994–2009	
USGS 18	Tritium	20	0	1977–2009	
	Strontium-90	19	0	1990–2009	
USGS 19	Tritium	80	0	1966–2009	
	Cesium-137	24	0	1991–2009	
	Gross alpha	23	0	1994–2009	
	Gross beta	23	0	1994–2009	
USGS 22	Tritium	59	0	1972–2009	
USGS 23	Tritium	28	0	1977–2009	
	Cesium-137	24	0	1991–2009	
	Gross alpha	23	0	1994–2009	
	Gross beta	23	0	1994–2009	
USGS 26	Tritium	28	0	1977–2009	
	Strontium-90	26	0	1991–2009	
	Cesium-137	24	0	1991–2009	
	Gross alpha	23	0	1994–2009	
	Gross beta	23	0	1994–2009	
USGS 27	Tritium	80	0	1967–2009	
	Cesium-137	24	0	1991–2009	
	Gross alpha	23	0	1994–2009	
	Gross beta	23	0	1994–2009	
USGS 29	Tritium	22	0	1977–2009	
	Strontium-90	18	0	1991–2009	
USGS 31	Tritium	22	0	1977–2009	
	Strontium-90	18	0	1991–2009	
USGS 32	Tritium	22	0	1977–2009	
	Strontium-90	18	0	1991–2009	
USGS 83	Tritium	95	1	1966–2009	9,000 ±2,000 in 1968
	Cesium-137	22	0	1994–2009	
	Gross alpha	22	0	1994–2009	
	Gross beta	22	0	1994–2009	
USGS 86	Tritium	79	0	1967–2009	
	Cesium-137	28	0	1980–2009	
	Gross alpha	27	0	1971–2009	
	Gross beta	27	0	1971–2009	

Table 6.   Summary of water-quality results for selected radiochemical constituents in water in water from wells and surface-water sites at and near the Idaho National Laboratory, Idaho.—Continued

[Data are available at http://waterdata.usgs.gov/nwis. **Local name:** Local well identifier used in this study. BLR, Big Lost River; BC, Birch Creek; LLR, Little Lost River. **Constituent:** Concentrations are in picocuries per liter. **Remarks:** Concentrations are given for values greater than the reporting level. Uncertainty is sample standard deviation]

Local name	Constituent	Number of samples	Number greater than reporting level	Period of record	Remarks
			Wells—Continued		
USGS 97	Tritium	98	3	1973–2009	800 ±200 in 1978; 600 ±200 in 1978; and 900 ±200 in 1979
	Strontium-90	33	0	1973–2009	
	Cesium-137	27	0	1994–2009	
	Gross alpha	23	0	1994–2009	
	Gross beta	23	2	1994–2009	6 ±2 in 1996; 10 ±2 in 1999
	Americium-241	23	0	1994–2009	
	Plutonium-238	23	0	1994–2009	
	Plutonium-239,240	23	0	1994–2009	
USGS 98	Tritium	93	0	1973–2009	
	Strontium-90	25	0	1973–2009	
	Cesium-137	30	0	1973–2009	
	Gross alpha	22	0	1994–2009	
	Gross beta	22	1	1994–2009	6 ±2 in 1995
	Americium-241	24	0	1989–2009	
	Plutonium-238	24	0	1989–2009	
	Plutonium-239,240	24	0	1989–2009	
USGS 99	Tritium	80	1	1974–2009	600 ±200 in 1984
USGS 100	Tritium	90	1	1974–2009	4,100 ±500 in 1978
USGS 101	Tritium	65	0	1974–2009	
	Cesium-137	23	0	1991–2009	
	Gross alpha	23	0	1994–2009	
	Gross beta	23	1	1994–2009	7 ±2 in 1999
USGS 102	Tritium	17	0	1994–2009	
	Strontium-90	15	0	1995–2009	
USGS 103	Tritium	104	2	1980–2005	800 ±200 in 1983; 1,200 ±300 in 1985
	Cesium-137	21	0	1994–2005	
	Gross alpha	21	0	1994–2005	
	Gross beta	21	2	1994–2005	10 ±2 in 1995; 7 ±2 in 1996
USGS 107	Tritium	68	0	1980–2009	
	Cesium-137	23	0	1994–2009	
	Gross alpha	24	0	1994–2009	
	Gross beta	24	5	1994–2009	6 ±2 in 1995; 1997; 1999; 7 ±2 in 2000; 11 ±2 in 2007
USGS 109	Tritium	55	0	1980–2009	
	Cesium-137	25	0	1987–2009	
	Gross alpha	24	0	1994–2009	
	Gross beta	24	4	1994–2009	6 ±2 in 1994, 1996, 1999; 7 ±2 in 1999
USGS 110	Tritium	41	0	1980–1992	
USGS 110A	Tritium	20	0	1995–2009	
	Cesium-137	20	0	1995–2009	
	Gross alpha	20	0	1995–2009	
	Gross beta	20	3	1995–2009	10 ±2 in 1995; 7 ±2 in 1996; 6 ±2 in 2006
USGS 117	Tritium	65	0	1987–2009	
	Strontium-90	64	0	1987–2009	
	Cesium-137	39	0	1987–2009	
	Americium-241	39	0	1987–2009	
	Plutonium-238	39	0	1987–2009	
	Plutonium-239,240	39	0	1987–2009	

Table 6   65

Table 6. Summary of water-quality results for selected radiochemical constituents in water in water from wells and surface-water sites at and near the Idaho National Laboratory, Idaho.—Continued

[Data are available at http://waterdata.usgs.gov/nwis. **Local name:** Local well identifier used in this study. BLR, Big Lost River; BC, Birch Creek; LLR, Little Lost River. **Constituent:** Concentrations are in picocuries per liter. **Remarks:** Concentrations are given for values greater than the reporting level. Uncertainty is sample standard deviation]

Local name	Constituent	Number of samples	Number greater than reporting level	Period of record	Remarks
		Wells—Continued			
USGS 119	Tritium	69	0	1987–2009	
	Strontium-90	68	0	1987–2009	
	Cesium-137	45	0	1987–2009	
	Americium-241	44	0	1987–2009	
	Plutonium-238	44	0	1987–2009	
	Plutonium-239,240	44	0	1987–2009	
USGS 121	Tritium	32	0	1991–2009	
	Strontium-90	32	0	1991–2009	
USGS 125	Tritium	25	0	1995–2009	
	Cesium-137	23	0	1995–2009	
	Gross alpha	23	0	1995–2009	
	Gross beta	23	2	1995–2009	6 ±2 in 1995; 1999
USGS 126 A	Tritium	13	0	2000–2009	
	Cesium-137	12	0	2000–2009	
	Gross alpha	12	0	2000–2009	
	Gross beta	12	0	2000–2009	
USGS 126 B	Tritium	13	0	2000–2009	
	Cesium-137	12	0	2000–2009	
	Gross alpha	12	0	2000–2009	
	Gross beta	12	0	2000–2009	
WS INEL-1	Tritium	50	0	1979–2009	
		Surface-water sites			
Big Lost River at Experimental Dairy Farm, near Howe	Tritium	28	0	1976–2009	
	Cesium-137	27	0	1980–2009	
	Gross alpha	26	1	1976–2009	4.8 ±1.5 in 1995
	Gross beta	26	2	1976–2009	15 ±2 in 1995; 2.5 ±0.8 in 1998
Big Lost River below INL diversion, near Arco	Tritium	29	0	1976–2009	
	Cesium-137	27	0	1980–2009	
	Gross alpha	23	0	1976–2009	
	Gross beta	23	1	1976–2009	13 ±2 in 1995
Big Lost River below Mackay Reservoir, near Mackay	Tritium	33	0	1990–2009	
	Cesium-137	32	0	1990–2009	
	Gross alpha	30	0	1990–2009	
	Gross beta	30	0	1990–2009	
Big Lost River near Arco	Tritium	37	0	1976–2009	
	Cesium-137	33	0	1979–2009	
	Gross alpha	31	0	1976–2009	
	Gross beta	31	3	1976–2009	6 ±2 in 1982, 2000; 14 ±2 in 1995
Birch Creek at Blue Dome, near Reno	Tritium	61	0	1972–2009	
Little Lost River near Howe	Tritium	69	0	1972–2009	
Mud Lake near Terreton	Tritium	70	0	1970–2009	

Appendixes

Appendix files are available for download at http://pubs.usgs.gov/sir/2012/5169.

Appendix A. Package 'Trends'

Appendix B. Plots of pH, specific conductance, and water temperature from selected sites, 1950–2009.

Appendix C. Plots of selected constituents from selected sites, 1949–2009.

Appendix D. Theil-Sen slope line and p-values for selected constituents for uncensored data from selected sites.

Appendix E. Theil-Sen slope line and p-values for selected constituents for censored data from selected sites.

www.ingramcontent.com/pod-product-compliance
Lightning Source LLC
Chambersburg PA
CBHW081604170526
45166CB00009B/2813